High Up, Rolling Sea, Deep Down, Ice Cold

An Ocean Geologist Explores the World

Praise from Early Readers

As a former merchant mariner, engineering officer, I can tell you that Peter's description of life at sea is true to my experiences. It's a life distant from landlubbers, but Peter invites you to see and feel it. His telling brought back good, bad, and ugly memories. Here's one good one: Whether it is the sunrises, sunsets, watching whales and porpoises, a sky full of constellations, and sometimes seeing the North Star and Southern Cross in the same night sky. Scientists and crew form a kinship in their appreciation of their uncommon workplace.

—Jim Borell, retired marine engineer

Now retired, I was an oceanographer (geologic) for 33 years and had been to sea numerous times. For the first seven years of my career, both Peter Vogt and I were part of a Naval Oceanographic Office research group called GOFAR. So, it is not surprising that I would fully enjoy reading his book as well as experience many déjà vu moments.

Peter has combined his considerable talents to produce a larger-than-life, frequently humorous, narrative about his travels and experiences as an oceanographer (specifically, an ocean geologist/geophysicist). It is an engaging story that has it all: adventure, hardship, exploration, discovery, and fulfilment. He supplements his narrative by including numerous photographs as well as correspondence between him and his wife while away from home.

—Frederick Bowles, PhD

Peter invites us along as he camps out on polar ice, hops across continents, explores underwater volcanoes, and maps the sea floor. His knowledge is extensive, his experience vast, and his curiosity about the natural world unquenchable. Peppered with amusing anecdotes and thoughtful letters home to Randi, his beloved wife, this is a remarkable journey of a restless man questing for answers, a quest that he continues to this day.

—Sherrod Sturrock, Director Emerita: Calvert Marine Museum

high up
rolling sea
deep down
ice cold

 cologist Explores the Worl

B. Vogt

High Up, Rolling Sea, Deep Down, Ice Cold:
An Ocean Geologist Explores the World
by Peter R. Vogt

Editors
William Lambrecht and Sandra Olivetti Martin
New Bay Books
Fairhaven, Maryland
NewBayBooks@gmail.com

Cover by Suzanne Shelden

Interior design by Suzanne Shelden
Shelden Studios
Prince Frederick, Maryland
sheldenstudios@comcast.net

A Note on Type:
Cover and section heads are set in Bullion WF
and text is set in Garamond Premier Pro

Library of Congress
Cataloging-in-Publication Data

ISBN: 979-8-9882998-7-5
Printed in the United States of America
First Edition

Dedication

This opus is dedicated to my wife Randi,
née Randi Karen Stampen, who:

Met my icebreaker *Atka* standing on a Bergen dock
in 1966 and accepted my marriage proposal
on a nearby mountain the next day;

Saw me off on the research ship *Lynch* in 1970 at the Navy
Yard before Hurricane Agnes silted up the Potomac channel;

Saw me off in 1969 at Andrews AFB on a five-week airborne
mission and forgave me for taking the car keys, leaving her
stranded with infant son Erik;

Drove to Norfolk in 1975 with two young sons
to welcome back their pop, who was waving
from the drilling derrick of the good ship
Glomar Challenger;

Saved my many letters and postcards home
from distant ports, making this book possible;

Lastly and most importantly, put up
with my many absences and cared for our sons,
even over the Christmas holidays.

Foreword

You are in for a treat when you read this illuminating book. Dr. Peter Vogt has brilliantly shown what it is like to search for and find the wonderful clues about our planet's origin and evolution. From a youth inspired by the beauty of ice—from glaciers in Austria and Antarctica, to the seemingly endless ice-covered polar seas, Dr. Vogt pulls us along on the discoveries. He has the curiosity of the best of our species, as an intrepid searcher and documenter of Earth's physical wonders.

Beyond his early infatuation with ice, he was lured into to the study of the ocean, the most significant region of our planet's surface. Through relentless seas, hot and cold, sickness and health, endless voyages to stunning vistas, his experiences in studying the changing nature of Earth are a testament to the value and role that historical, real-life experiences play with our interpretations of the science of discovery.

If this book had been written in the 1960s, I am sure I would have read it when I was a young teenager, so hungry was I to devour stories of adventure and exploration of the Earth. Today, after I myself have spent more than four years at sea, I am humbled by the amount of time that Dr. Vogt spent on these voyages, from month after month, separated from his family, and having to depend on letters to keep him attached, through thick and thin. This could not have been easy. I imagine his family also suffered because of the unending time spent apart.

Dr. Vogt did his best to also understand his international colleagues by learning how to speak many languages. This is an attribute very rarely appreciated by the American people, but always appreciated by shipmates around the world. Indeed, the international programs in which Dr. Vogt participated were greatly strengthened by his ability to communicate, a quality other American scientists should emulate. Dr. Vogt also became an unwitting ambassador from the U.S. Naval Research Laboratory to the American academic-science community. I too benefited from this connection in the 1990s when I was invited by Dr. Vogt to lead a program on the changes of the Arctic environment at the Naval Research Laboratory.

I also recommend this book highly to any and all who are concerned about the huge changes our Earth is undergoing due to the carelessness of our fellow humans on the environment of our only planet. We have a long way to go.

—Kathleen Crane, PhD
Author: *Sea Legs*

Contents

In Antarctica, Peter ran this tracked vehicle, called a Nodwell, into soft ice.
He had to be pulled out by, yes, another Nodwell.

Introduction

WHAT IS THIS LONG BOOK IN YOUR HANDS?

Is this a travelogue featuring exotic ports and ocean islands? A 50-year history of ocean-floor geology research? A geology primer? Life on research ships on the high seas? Diving in both sub and submersible? About mountain glaciers, Arctic pack ice, and Antarctica? Being a scientist at a US Naval laboratory vs in academia? Whales and fish, storms, and sunsets? Buzzings by Soviet Badger bombers and tailings by warships during the Cold War? Searching for toxic waste dumped into the sea?

This book is all of the above, and more.

OCEAN GEOLOGY HAS BEEN MY CALLING

Ocean geology was the genesis of this book because most of my career adventures were on ships, planes, even a submarine and two submersibles, learning about the ocean floor (its shape, properties, and processes) and what lies below, most commonly a layer of sediment on hard basaltic oceanic crust. Much of what we studied related to plate tectonics, a paradigm unknown when I was an undergrad. 'My' research included the earliest phases of Atlantic opening, when North America and Africa began to separate about 175 million years ago, early in the age of dinosaurs. (I put quotes around 'my' because this was all team effort.)

Our expeditions also discovered and explored active mud volcanoes, sub-marine landslides, seafloor grooves plowed by deep iceberg keels of past ice ages, and mostly ancient sub-marine volcanoes called seamounts, the highest ones sunken former islands whose island tops had been eroded off. The geoscience cruises reported in this book explored all the world's oceans except the Roaring Forties, an exception for which my stomach is forever grateful.

(Seafarers who began sailing from Europe to the Orient around South America's Cape Horn or South Africa's Cape Agulhas or Cape of Good Hope—or sailing f rom the US East Coast around South America to California before the Panama Canal—encountered the stormiest seas on the planet generally between 40 South latitude and 50 S, hence the term Roaring Forties. In those latitudes there is no land to slow winds.)

I even circumnavigated (1968) the Southern Hemisphere (over land and sea) as science watch-stander aboard the Project Magnet's special Super Constellation. Was this about geology? Yes, but about Earth's core, which produces the changing geomagnetic field that turns compass needles and aids—or long aided—navigators. Project Magnet and their SuperConnie are long gone—the 'history' part of this book.

The world's oceans are large. Most of the research expeditions reported in this book explored remote areas. We often went days or even weeks without sighting another vessel. Working in or crossing traveled seas—trade routes and in-conflict, fishing, tourist-cruising, and gas/oil areas—we were often approached by some curious mate or skipper for a look-see. What kind of ship were we, and why stopped on the high seas?

In many ways, my most memorable ocean geology experience was as co-chief scientist on Leg 43 on the *Glomar Challenger* (1975). As a curious coincidence, we cored holes in the fictitious Bermuda Triangle. In an exciting epilogue,

I almost 'kidnapped' the Soviet researcher on that cruise—giving him the joy of visiting beautiful Southern Maryland and Washington DC.

SIDE GIGS

Not all my career adventures were geological, and some were on land or even in the office. My first ocean research cruise (1958) tracked the Equatorial Countercurrent (Eastern Pacific), not the sediments and rock far below. In summer 1959, I worked for a mining company (Phelps-Dodge) prospecting for ores in the blazing hot Arizona desert. College-age climbs of glaciated volcanoes in Mexico (1960) and South America (1963) were far from any ocean, as was the Hintereisferner glacier in the Austrian Alps (1962). As a research assistant in Antarctica, I worked in ice-free areas studying permafrost fractures. My first ocean geology cruise as chief scientist (1970) was diverted to check if a scuttled Army ship loaded with nerve gas had broken up in descent, the gas leaking from canisters.

In 1979–'80 I was chosen as a member of the Naval Research Lab team to refute the claim by the Ruina Panel that the Vela satellite double flash was a natural event, not the detection of an undeclared nuclear bomb test by Israel with South African help in the remote southwestern Indian Ocean. Most historians agree with NRL's evidence that this was a cover-up by the Carter White House. Science at the service of geopolitics! That was as much adrenaline rush for me as any research cruise or flight.

In 1993, I led a multi-agency team to investigate the USSR's many radioactive waste dumps in the Kara Sea. I had flown to Moscow earlier with a colleague from the Office of Naval Research to hire a Russian ship. Our team came to Kiel, Germany to await the *Mendeleev*, which arrived days late. But a disgruntled Soviet—now Russian—admiral killed the agreed-upon plan, so we took off our instruments and went home.

AS TIME GOES BY

This book also serves as a half-century history: Technology, the plate tectonic revolution, climate change, gender equality at sea.

Much had changed between my first (1958) and last (2010) ocean-research cruises. From the 1950s into the '70s, the seamen, officers, students, and senior scientists were almost all White males. The USSR was ahead with respect to women—as I discovered on the 1975 expedition. A few women also served on my *Glomar Challenger* expedition, albeit none on the science team. Only White males, Navy and civilian, on my 1968, 1969, 1984, and 1990 aircraft deployments—but 1990 was 35 years ago.

By the late '80s and certainly today, male dominance on research ships was much reduced. Today's male student watch standers likely have female thesis advisors or chief scientists. There were no women on my Coast Guard (1965) and Navy (1966) icebreaker cruises. That had changed by the time of my 2002 USCGC *Healy* icebreaker expedition. Progress in female representation on research ships probably depends on the nation and institution. Any reader interested in this aspect should also read *Sea Legs* by my fellow marine geologist Kathy Crane.

The first half of my career featured the US-USSR Cold War. On November 9, 1989, I disembarked from the Norwegian research ship *Haakon Mosby*, checked into a Bergen hotel, turned on the TV and learned the Berlin Wall was being peacefully breached. The ensuing breakup of the USSR changed US Naval funding priorities—reducing the research focus on anti-submarine warfare. The increased parking space at NRL paralleled reduced funding.

Satellite navigation was in the early stages when I first shipped out, and officers still often used sextants. There was no email yet, and we sent and received paper letters by way of embassies. The radio operator posted a paper printout of the day's news in the galley. Calling home meant patching a call via a ham operator, standing in the radio room with others listening, and teaching the other end to say 'over' but nothing personal. Booze restrictions have become more strict, maybe with the exception of Russian ships. There are no more swim calls, even with an armed seaman on the flying bridge with a rifle ready to shoot sharks. Dumping garbage or trash overboard is prohibited by international regulation. There are surely still Neptune's Day ceremonies, but hazing might be more constrained by law.

However, other things have scarcely changed, for example ship construction and diesel-electric propulsion. Watch schedules, e.g., four hours on, eight off, remain the way they've been for centuries. The chief cook may be the most important person on board. Chief scientists will probably always argue with skippers about when to break off research and set course for the next port.

Ship operation contingents have become smaller, and many of today's chief engineers monitor their hot engine rooms remotely from their cabins. Skippers have their traditional reserved seats in the wheelhouse. Identity and warning flags and lights likely remain unchanged. That continuity is important for research ships underway towing sensors on cables or on station with instruments or sampling gear hanging down on long cables. Modern research ships all have bow thrusters to keep the vessel in one spot. Radar is as important as 50 years ago, but today's monitors are all digital with colored screens.

Towed and autonomous (free-swimming AUVs) instrument packages are now sophisticated, and once launched from a ship, do most of the data and sample collection, plus high-resolution imaging of particular ocean floor targets, such as hydrothermal vents. Even manned submersibles-like the Russian Mirs I dove on (1998) are less often used and cost-effective for research today (much as robot landers can do most of what astronauts could on the Moon), and far cheaper. Today's computers are speedboats compared to the dugout canoes I programmed 60 years ago, feeding punch cards with Fortran programs into an IBM 1620 or leaving a card deck at a computing center in the afternoon for a cheaper overnight run.

In my field of marine geoscience, the biggest change happened mostly between 1958 and 1968, the year most associated with the plate tectonic revolution, a paradigm shift second to what Charles Darwin started a century earlier. In my Caltech geology and geophysics courses (1958–1960), most geologists wrote off continental drift as a fairy tale and rarely if ever mentioned it. Caltech had leading scientists, and I had courses from Charlie Richter of the Richter Scale; Frank Press, later President Jimmy

Carter's science advisor; field geologist Robert Sharp; and others. One of my Caltech student field trips was to the infamous San Andreas fault. Scarcely a decade later this great fault would be recognized as part of the boundary between the giant Pacific and North American plates. By chance or choice, the American Geophysical Union holds its week-long annual meetings—I attended and gave lectures at many—in San Francisco, which had been demolished in a huge 1908 earthquake on this fault system.

Most of the career-adventure chapters in this book deal directly or indirectly with the oceanic side of plate tectonics and this history of plate motions, in the central Atlantic back more than 150 million years. Mapping and interpreting the linear magnetic anomalies over oceanic crust—a combined record of past sea-floor spreading from the Mid-Oceanic ridge and polarity flips of the Earth's magnetic field, which turns compass needles. It was my good luck to be hired by the Washington, DC, US Naval Oceanographic Office (Washington DC area) in 1967, just in time to analyze, with my newfound colleagues of their Magnetic Division, their large magnetic data holdings—data whose meaning no one had known.

Upon joining a new NAVOCEANO research group housed at Randle Cliffs, the growing Vogt family moved to Calvert County and we elders remain here to this day. In 1975, I transferred to the Naval Research Laboratory and began to commute to DC, an hour each way. My location got me interested in the geology of Southern Maryland, the Calvert Cliffs and Chesapeake Bay: brand-new research projects before and after retirement.

NAVOCEANO, notwithstanding its pre-academic 19th century ancestry, was later absorbed into other agencies and no longer exists. My only other extinct employer was the Phelps-Dodge Mining Company, for which I had worked in 1959.

The concept of ongoing climate change was not mentioned in any Caltech lecture. It so happened, by chance, that atmospheric carbon dioxide was first measured by David Keeling of Scripps on Mauna Loa when I was a Caltech sophomore. Keeling, whom I likely never met, measured 313 parts per million in March, 1958, a few months before I shipped out on the Scripps' *Horizon*. Of course, he could not have foreseen the future increase. In February 2025—as this book was being written—carbon dioxide stood at 427 ppm, >50 percent higher than pre-Industrial levels, and probably the highest in at least three million years. The Austrian glacier Hintereisferner, on which I and other University of Innsbruck students helped our glaciology professor excavate a sampling pit in 1962, has lost mass every year since 1980. The glaciers we climbed on the Mexican volcano Popocatepetl in 1960 are gone. The world's highest ski slopes (Chacaltaya in Bolivia) we visited in 1963 are now bare rock and abandoned. The average ice cover season on Lake Mendota, on which I skated while a graduate student at the University of Wisconsin (1963–'67), is 20 days shorter today. In our Chesapeake region, we have not seen any more winters as cold and icy as those in 1976–'82, more than four decades ago.

Peter having an "ice liberty" beer while standing on
the Arctic Ocean, with USCGC *Healy* in background.

During Ice Liberty, photographed on an ice floe six feet thick in the 4,000 feet-plus deep Arctic Ocean, well north of Alaska. At 16,000 tons, USCGC *Healy* is the biggest ship I ever traveled on and one of my last expeditions.

On my two icebreaker cruises (1965 and 1966), we had no problems finding thick, large ice floes for ice liberty, measuring gravity, and getting stuck. When I arrived in Barrow, Alaska, to board the icebreaker *Healy* in 2002, I was surprised that no sea ice was to be seen, and surf breaking on the shore eroded the coastal road. We steamed for more than a day before reaching the pack, and later had a hard time finding a floe big and thick enough for ice-anchoring for ice liberty.

ACADEMIC AND NAVAL SCIENCE:
ONE FOOT IN EACH WORLD

Most basic ocean research is done in universities or institutions like Woods Hole Oceanographic Institution, Scripps

Institution of Oceanography, or Lamont-Doherty Earth Observatory. These places are affiliated with or parts of universities and host graduate students doing thesis work, but no undergrads. I have collaborated or otherwise interacted with geoscientists at these institutes and others, mostly with Dr. Brian Tucholke of Woods Hole. Among universities, Princeton stands out for me in the 1970s, notably Prof. W. Jason Morgan, who formulated plate tectonics. While many such researchers have been in part funded by grants from the Office of Naval Research, they are not otherwise affiliated or employed by the US Navy.

My career situation was unique. I was proud and privileged to be employed, for 37 years, as a civilian scientist (marine geophysics), first by the US Naval Oceanographic Office (NAVOCEANO) from 1967 to 1975, then until 2004 by the Naval Research Laboratory. The global and high-tech nature of the US Navy is founded on R&D at all levels, starting with Basic (called 6.1). This category does not include all important basic research, but only that which could in the future lead to newer, better, or better-functioning Naval operations. Knowing water depth and seafloor physical properties and processes brings together ocean geology and Naval applications.

The D is for development, and the path to applications is numbered 6.2 to 6.6. The Naval Research Laboratory, founded in 1923 from a proposal by Thomas Edison, does more types and levels of science than any other military laboratory in the world. Unlike my research peers from academia, I also did applied, including classified, research.

We checked and initialed our Branch safes at day's end— though it would not have occurred to me nor any NRL colleagues to take home classified materials. Yes, I worked on basic research at home and at NRL after hours, preparing for a paper to be presented at a research workshop or conference, for example of the American Geophysical Union. There I chatted with academic colleagues and learned about what they had found. Anything of possible Naval interest I would report back. That was one reason NRL paid me to attend and encouraged me to publish in peer-reviewed research journals. NRL expected its basic research to be competitive with academia. Our Branch 6.1 program was regularly reviewed and rated by an outside panel, including academics. I served as the primary 6.1 scientist in our Branch, organizing and leading our defense.

I often interacted with the uniformed Navy: briefing White Hats at NRL, as scientist observer on the unique NR-1 sub (1999); as watch-stander or scientist on Navy aircraft flown by Navy pilots and air-crews (1968, 1969, 1984, 1990); and accepting research awards from NAVOCEANO and NRL Commanding Officers (captain rank). While I was never in uniform, my wallet once carried an ID card telling possible enemy captors to treat me as of captain rank. On the UK–US airbase on Diego Garcia atoll, which lacks commercial lodging, I was assigned to the Bass Operation Quarters. My two academia colleagues, I recall, were assigned to enlisted quarters—although on that project they initiated the project and headed the science team.

Even before civil service, University of Wisconsin grad student Vogt collected thesis data from Navy icebreaker USS *Atka* (1966) and worked in Antarctica back when the US Navy handled long-distance logistics. The McMurdo outhouse was a long row with no partitions, so at one time or another I likely sat next to a Navy pilot or crew member, but I don't recall they were in uniform.

Interactions with the US Coast Guard—of which I have a high opinion—occurred on 1965 and 2002 icebreaker cruises. Interactions with the foreign military were at a distance and uncivil. Our 1964 Woods Hole landing on Zebirget Island, Red Sea, and geological sampling was under binocular eyes from Egyptian gunboats. The Cold War *Northwind* and *Atka* icebreaker cruises were repeatedly buzzed by Soviet Badger aircraft—some of which dropped red flares—and tailed variously by trawler, destroyer and missile frigate.

SCIENTISTS AS TOURISTS

And life on a research ship is not all about standing watches in the lab, winching in a cable or emptying a dredge of rocks. Off watch there are nightly movies or card games. When whales, strange ships, mysterious shores, or polar bears are sighted, those off watch—scientist or seaman—gather on deck, binoculars and cameras in hand. In the tropics, weather permitting, we watched sunsets hoping to catch the brief green flash. The Neptune's Day Equator crossing hazing and ceremonies (on *Chain*, Indian Ocean, 1963, and on the Soviet *Kurchatov*, Atlantic, 1975) were memorable if a bit uncomfortable experiences.

Port stops—to refuel, restock provisions, or collect a spare part—offer some free time to explore port cities or, alas, only to get plastered in harbor bars. As a mere grad student watch stander, I had more shore leave in ports than later as senior or chief scientist. Then I sometimes showed the flag, staying on board to offer locals tours of the ship. Most visitors were in truth not really interested in our magnetometers or gravity meters. Soviet research ships offered caviar and vodka, attracting more visitors.

Free time ranged from hours to days, as in La Spezia (Italy), where the Woods Hole ship *Chain* (1964) was moored for days waiting for a vital spare part to be flown in. I explored other Italian towns by train. When the icebreaker *Northwind* (1965) damaged a screw in heavy Arctic ice and spent many days dry-docked in Newcastle (England), I explored Scotland, catching the Edinburgh Tattoo and hiking up Ben Nevis.

The most exciting stops were not port stops but going ashore with land geologists on remote ocean islands: American geologists on a desert island in the Red Sea (1964) and Russian geologists on St Helena and Tristan da Cunha in the South Atlantic (1975).

LUCK AND LOGIC OF THE DRAW

How did a boy from Dayton, Ohio, wind up in such places as Tristan da Cunha, La Spezia, Italy, the Suez Canal, and Antarctica?

Along with my lifelong interests in nature and travel, I was good in high school STEM subjects and so pursued science in college, where my interests turned from physics to geoscience. After high school and throughout my life, whenever opportunities for global research and travel knocked, I opened those doors. Once I achieved my PhD, I had to earn a living and by 1967 to help support a family; I did so for 37 years as a full-time scientist in US Naval research. Finally luck and my supportive wife played major roles in my career, which I relive in this book.

PART I

Ice and Snow: A Lifelong Infatuation

Of the eight billions currently on planet Earth, the majority have likely never seen nor felt real snow and ice. Of those familiar with the stuff, the most are likely not too fond, and if they have resources, escape winters for warmer climes. Those fond of snow are likely only found when they can ski on it. Of course many children like to play in the snow, as do many dogs.

I am an exception. Snow and ice, including glaciers, are my lifelong fascinations.

I spent my boyhood in Ohio, living right next to a small lake. Like other kids, I welcomed snow days when school closed. Like other parents, mine were not so thrilled. For us kids, snow meant snowmen, sledding, snow forts and snowball fights. That doesn't make me unusual. A little more unusual, my fascination was more with ice than snow, though both are just the solid form of water.

When the lake froze, we all went ice skating. I suppose that's when my ice infatuation began. The newly safe ice—before snow and sleet mostly ruined it—looked black, but cracks decorated it in gracefully splaying filaments and curtains. Breathing muskrats or methane gas escaping from the mud created stacks of white bubbles which froze into the thickening ice. The deepest bubble, right under the ice, was always translucent, a clear measure of ice thickness and thus safety.

New lake ice was safe for people once about three to four inches thick. Not quite that thick, and it would crack but not break under one's skate feet. Yes, I broke through once, but the water was only waist deep and not far from shore.

Once safely thick, it would still crack—on sub zero days— not due to skaters but to thermal contraction. Ice, like most other solids, shrinks as it cools. The cracking sounds made exciting twangs, like a giant plucking a giant bass string. The ice music scared some skaters who did not understand.

Kids throwing rocks and sticks on the ice before it was safe to skate on left hazards, especially for those like me who liked to skate backwards. An occasional frozen dead fish and leaves blown onto the ice created more things to skate around. I switched from figure to hockey skates for extra speed and less tripping on rocks or sticks frozen into the ice.

As lake ice thaws and refreezes repeatedly in spring, it recrystallizes into 'candle ice': lots of vertical cheek-to-jowl 'candles.' Candle ice has little flexural strength and even when relatively thick becomes easy to break through.

I always got sad in spring when frozen lakes and rivers thawed and snow melted. I even changed the lyrics of the old New Christy Minstrel hit "Green Grass." In my version, the lyrics go:

> *Snow, Snow the snow lies still,*
> *on the north side of the hill...*
>
> *Snow, snow, I'm going away*
> *to where the air is colder still.*

CHAPTER 1

Searching for Snow and Ice in Sunny California

Once our family had moved to Southern California, ice skating became restricted to rinks. You skated around and around but got nowhere—kind of like most lives. Yes, there was music and no sticks or rocks or frozen fish. However, rink ice became chopped up.

I learned to skate backwards to rock beats by imitating other skaters. Instead of extending, even flailing to maintain balance, I held my arms and hands behind my back. I bought speed skates and had their ends grounded into a curve so I could skate backwards on those as well.

Snow was, of course, lacking, except in the mountains, so I did learn to ski, but never really well. Some of us from Santa Barbara High School would drive up into the nearby Coast Range on those rare winter days that snow fell at high altitudes and come back to town with trunks full of the white stuff.

I did have one memorable outdoor skating experience in California. My brother Volker and I planned to ski at Mammoth Lakes, but that Christmas vacation there was no snow. So we took along our skates.

The High Sierra is laced with many conifer-ringed lakes, carved by Ice Age glaciers. The road to Mammoth went right along one. The ice was like glass, and not a soul out on it. We parked and ventured out on foot, perfectly safe. Once we were

3

skating, some others soon parked and joined us. The grand vista was dominated by Mount Ritter and the Minarets, the jagged peaks along the Ritter Range. We forgot about the lack of snow.

Chapter 2

Climbing Glacier Ice South of the Border

To experience the real thing—actual glaciers—we headed south of the border. It was late 1960. Three of us—my brother, Volker and I, plus Herb Rice, a fellow Caltech undergraduate—drove Herb's ancient, double-clutch Ford from sea-level southern California to Mexico City.

Our lungs had not adjusted to the thin air, but Christmas vacation was only so long. Spending a night in a stone shelter at perhaps 13,000 feet elevation, we rose before dawn to climb the 17,694-foot summit of Popocatepetl (*smoking mountain* in Nahuatl). Popo was clearly visible from Mexico City when city smog had been blown away and was frequently active; it erupted in December 2018.

The Volcano Popocatepetl

We proceeded onto the steeply sloping ice, wearing crampons and likely carrying ice axes. I had no pains and recall feeling superior, well above and thus ahead of the others.

But suddenly, a pounding headache and sluggish breathing slowed me down. Only one step per breath. I was the last on the summit—a spot on the crater rim. Exhausted and inhaling sulfurous crater fumes, I collapsed in the snow. My brother's prodding finally got me up for the descent. I'm sure glad he didn't leave me there to expire.

Back then, Popo had a few glaciers, but by 2001 they were gone, thanks to some combination of volcanism and climate change. The stratovolcano is still capped with snow, but there is not enough ice to move downslope and spot telltale crevasses.

Our Mexican mountain adventure did not end with Popo-catepetl. Our high point at 18,491 feet—the third highest peak in North America—was Pico de Orizaba (*Citlatepetl* in Nahuatl). We left our car in a town two miles in elevation below the volcano summit and hired a Mexican guide and his mule to shlep our sleeping bags up to about 14,000 feet, where we took shelter in a lava cave. It was Christmas Eve. The cave floor was covered with straw, and the mule stayed just outside the cave. As close to a nativity scene as I would ever experience.

We rose before dawn to experience the view from the top before afternoon clouds shrouded the summit. Our guide had made some special tea he claimed was good against altitude sickness. Indeed we had no big problems, and from Orizaba's icy slope we looked east to see the cloud deck on coastal plain jungle and the gleam of the Gulf of Mexico more than three miles below us.

Orizaba had some nine small glaciers, but we ascended on the unglaciated side, where intense tropical sunshine had created *hielo pendiente,* a scalloped ice surface with reliefs up to several feet. A bit laborious to climb, but no crevasses to fall into and no danger of sliding down that long slope out of control to your doom. We did not tarry long on the summit, and had no sulfurous fumes. This stratovolcano had been dormant for around 1,500 years. Heading down and down, we inhaled ever more of that wonderful gas we call oxygen.

We continued past the cave where our guide awaited us and made it back to town (Ciudad Serdan, elev. 8,200 feet) by evening. It was the only time I have ever covered 4,000 feet up and then two miles vertical down in one very long day, on foot. We spent an unplanned but welcome extra day recuperating in town: A major festival—presumably Christmas—had our car trapped. Then we were on the road again.

Ever careful about what I drank and ate, I had avoided Montezuma's Revenge while in Mexico, but I must have been careless the last day.

There is little vegetation along that Texas highway just across the border. Welcome back to the United States.

Chapter 3

Onward from Caltech

That Christmas vacation mountain adventure was welcome relief from the four intellectually grueling years I spent at Caltech, then still all-male until graduate school. Caltech officials emphasized to us frosh that girls were not excluded because they were dumber than guys but because they too frequently dropped out to get married and have kids. The Institute, we were told, spent much more on us students than we paid in tuition.

Although starting out majoring in physics, I took some interesting geology courses exposing me directly in the outdoors and the fantastically diverse California geology. I got to visit the famous (and infamous) San Andreas Fault about a decade before it was recognized as boundary between the great Pacific and North America tectonic plates. I had gone backpacking in the Sierras and hiking in local coast ranges. So I switched majors to geophysics. Same classical physics as for physics majors but less of nuclear, quantum and other such modern stuff—and more geoscience.

The last year as a senior has always been a time to apply to graduate school. The University of Wisconsin offered coeds, beer, cold winter and the Polar Research Center housed at the former Brittingham estate outside Madison. Tired of application forms, I tossed forms from other schools. Indeed I was soon accepted. However, I had also been awarded an undergraduate Fulbright Scholarship at the University of Innsbruck for the academic year 1961–'62. Wisconsin was okay with me deferring grad school for that year.

CHAPTER 4

Innsbruck: Studies, Skis, and a Sesquipedalian Agglutinated Encounter

Leaving Caltech in 1961, I headed for Europe for an undergraduate Fulbright at the University of Innsbruck, Austria. I traveled from California to Austria first by train via Chicago to New York and then to Genoa, Italy, via ocean liner. Yes, there was already air travel, but many people still crossed the country by train and the Atlantic by boat. Italy was just where the liner stopped closest to Austria.

After all us Austrian Fulbrighters had been introduced to each other and to Austria in Vienna, I spent the summer traveling around Europe the way students did, and do.

Finally, I boarded the train to Innsbruck. Question: Why there? Are you, reader, surprised?

Answer: It was a place surrounded by snow-capped mountains and even real glaciers, at elevations of only a couple thousand meters. The Austrian Alps rise right above Innsbruck, which would host a Winter Olympics in 1976. The higher Alps, with glaciers I would traverse and help study, lay further south in the Ötztaler Alpen. That's also where some German hikers stumbled onto the melted-out mummy that would be named the Ice Man. (It was later determined by surveying that the mummy Ötzi was actually on the Italian side of the border. Sorry, Austria.)

In Innsbruck, I took courses in ice and glaciers decades before climate change and greenhouse effect became recognized by mainstream science and rational humanity. Other courses included physics and history. I soon dropped out of physics class due to declining eyesight. With myopia, I simply could not read the equations on the blackboard, and my technical German, although good, was not that good. Embarrassing for a Caltech alumnus who graduated with honors.

The history lectures were literally *Vorlesungen.* The professor actually read his lectures. I likely fell asleep and could always read the book. Back then there were no real final exams, just a one-on-one discussion with a key professor. You were adequate or excellent. I only got an adequate score, not due to poor German but freezing and flunking a simple question on planetary science. At least I was not trying for US credit from the courses.

Back then, courses were listed in catalogs as ST or CT. That's for Latin *sin tempore* meaning the lecture starts on time. CT is *cum tempore* meaning the lecture could begin up to 15 minutes later. I guess that gave some professors time for one last run down the slopes. After a major snow, the lecture halls were nearly empty. That was not at all about students being snowed in or a student demonstration.

I also owned skis, giant slalom ones which back then were as tall as my bent hands stretched above my head. They were much harder to turn than most modern short skis which, back then, were reserved for kids. The allegedly greatest and expert run was found on the upper slopes directly north of the town. I took the cable car up to the top and walked over to the top of the run. Without thinking long, I took the same gondola back down. Nuff said. I never advanced above an intermediate skier. By the way: At the bottom of that particular run, the ski patrol was positioned with first aid gear and evacuation sleds.

With another Fulbrighter named Anthony Oberdorfer, I roomed with an older landlady in downtown old Innsbruck, which basically was untouched by Allied WWII bombing. We had two beds but had to traipse outside on a long cold balcony to the toilet. I showered in the public bathhouse for a few Austrian shillings, which included a towel and soap. Many US students today are too soft and spoiled to live in such venues, but it was a lot more comfy than backpacking in the Sierras. I did not get along well with my roommate; I was kind of messy whilst he was neat and orderly, like the *Odd Couple* comedy years later.

Easter vacation 1962 was my chance to eat with the globally adored ski instructors and ski patrol. I was, of course, not one of them but was believed to be by the skiing public masses eating in the adjacent public cafeteria.

At a major ski area south of Innsbruck, I had been hired to replace the technician meteorologist so he could take his own Easter vacation. He was responsible for monitoring at regular intervals the meteorological instruments that were enclosed by a fence to keep the nosy skiing public away. He and the instruments were part of the *Lawinenforschungsdienst*. The English translation of that sesquipedalian agglutinated Germanic term is Avalanche Research Agency. If meteorological conditions such as temperature, humidity and snow accumulation increased avalanche risk to skiers, certain slopes might be closed. I was actually doing something that might save skiers from avalanches. Maybe it was the beginning of a career in ice and snow. I was also being admired by young Austrian ladies and envied by their boyfriends.

Alas, fate would ordain me to wait for that right lady. She was Norwegian-American, and it happened in Wisconsin five years later. Actually we had first noticed each other in 1964 on an outdoors club ski trip to Colorado. I had written her name down in my little book of girls to ask on dates, then lost the book and forgot her name. Apparently Cupid shot an arrow into the air he knew not where. However he restrung his bow and tried again a year later.

My first opportunity to research glaciers had to wait until summer, when winter snows had largely melted. I volunteered, or maybe was actually hired with other students, to assist University of Innsbruck glaciologist Herfried Hoinkes' work on the much-studied Hintereisferner, a small glacier in the Ötztaler Alps.

Carrying pickaxes, shovels and other tools, we trudged a long way single file up rocky moraines, then walked on skis to get onto the middle of the glacier. Hoinkes was studying the role of Sahara dust on glacier melting and ice budget. We helped excavate a deep pit several meters into the glacier surface so as to expose and sample a number of yellow snow layers, each layer a record of southerly winds blowing Sahara dust from desert dust storms across the Mediterranean, to be snowed or settled out onto the glacier. I'm sure this is related to glacier surface albedo (the proportion of the incident light reflected by a surface), which today more than ever is recognized as a factor in glacier retreat as both cause and consequence of climate change. Would glaciology be my future career?

As Innsbruck serves as a ski center in winter, the town in summer is a base for mountain climbing. The university offered beginner training both on glacier ice and on rock, and I signed up for one week of each. Our groups of a half dozen or so stayed in Alpine huts overnight. We learned to walk with crampons buckled to our boots and how to deploy our ice axes with lightning speed. Sunglasses were essential to protect against the sunny and icy glare.

Falling on a steep slope means plunging the pointed end of the ice ax into the ice to prevent sliding down the slope to a fatal high-speed collision with rocks or, perhaps, flying off a cliff. The shovel end of the ice ax is used to cut steps into the ice to enable the climb. Ice axes are always carried with a loop of rope around an arm. Woe to the climber who has to watch his ice ax slide down the mountain. I still have the ice ax today. Even in Southern Maryland, where I live, it can be pressed into service now and then.

My week of rock climbing began with a hike from our hut through an expanse of large boulders, some of them car-sized, parked there by melting glaciers long ago. We passed many boulders adorned with bronze memorial plaques with dates and names of climbers who had fallen to their death. Not exactly comforting. We learned how to rope ourselves together and began to climb.

I was in good shape then but had trouble hauling myself up. At one time I loudly cursed mountain climbing in German—*scheiss Bergsteigen* I announced (probably not good grammar for 'shit on mountain climbing'). But I made it up all those lower but not yet dangerous slopes. Farther up, we were inching along a narrow ledge. The important thing is to avoid what instinct suggests—leaning toward the cliff. We were now in the clouds and could only hear loosened rocks clattering their way down the cliff. As our guide explained, his white smiling teeth contrasting with a dark tan, that was just "Ludwig at work." I gathered Ludwig was the Austrian name for the wily mountain devil. We all made it to the summit, that rewarded us with bright sunshine and a grand view. Just enough time for some photos.

Summer adventures in the Austrian Alps.

Going down meant rappelling and using carabiners. We had been coached on that on an earlier climb. Rappelling is a fun and fast way to get down cliffs. One first has to overcome fear of stepping off a cliff backwards, and trust the rope and to what or whom it is tied.

I was now back in Innsbruck and getting ready to return to the States. Was I glad I had taken these two weeks' beginner training on ice and rock climbing? Yes. Would I ever want to do that again? Absolutely not!

CHAPTER 5

On the Big Ice of Antarctica

In late summer 1962, I boarded another passenger liner and returned to the US. After a brief visit to my parents in Santa Barbara, I headed back east for the US Antarctic Research Program orientation lectures in Skyland Hall, Shenandoah National Park.

This adventure involved real science: working for the austral summer 1962–'63 for the University of Wisconsin Geology Department's Robert Black and his research student, Tom Berg, as a field assistant on the Big Ice: Antarctica. Researching in the great outdoors was far more fun than labs and study halls.

To meet coeds, or any woman, I would have to wait a few more months. In Antarctica, I soon learned this truth: "there was a woman behind every tree."

Antarctica is almost totally buried under thick glacier ice, but Black and Berg were not studying ice, snow or glaciers. Small areas, including the Dry Valleys in the Royal Society Range, had been glaciated long ago, but today the valley floors sport patterned ground—the bare permafrost crisscrossed with ice-filled fractures, patterned as in dried mud puddles or crackle-finish china. The hexagons outlined by fractures were typically about 100 feet across. Black was studying the fractures to measure their rate of widening to find out when the glacier ice there disappeared.

Preparation for the Big Ice required a week or two USARP orientation at Skyland Resort in the Blue Ridge. Before that we had to get physical exams from our own doctors. This included eyesight and hearing. Alas, I flunked, barely, the Ishihara color test. But I successfully pleaded that on the Big Ice the colors were limited to blue sky, white snow and black rock.

At Skyland, a kind of comfy camp, we heard lectures by old-time explorers, learned to recognize frostbite on a buddy's nose and were issued official international-orange anoraks and parkas. These were emblazoned with blue USARP logos. I still have the anorak, somewhat bleached after 60 years, and often skied in it to earn admiration and detract from awkward skiing. Someone made off with the parka. Not sure if we were supposed to return this US government-issued garb.

Getting to the US main base at McMurdo—then as now—was mostly by way of Hawaii and Christchurch, New Zealand. Back then the military did the flying; in the Antarctic, the US Navy. Christchurch was, and maybe still is, called CheeChee by Americans heading to or from Antarctica. I recall it as kind of English, with lots of non-native northern fruit trees in bloom. The main river through town was the Avon. October was spring. We were supposed to stay there for a night or two to await our flight south to McMurdo. As it turned out, aircraft technical issues gave me nearly two weeks on New Zealand's South Island. No complaints from me about that unplanned vacation. I went hiking with a local outdoors club and took the train across the Southern Alps to the Tasman Sea side.

Finally we were on our way south, sitting along both sides of this large transport, already dressed ready for the cold. Through our starboard windows, flying ever south over the ice-covered ocean, we admired the 10,000–to–14,000-foot high peaks of the Royal Society Range. Through gaps among those peaks there stretched in the far distance the featureless white East Antarctic Ice Sheet. It's been there in some form for at least 33 million years and is likely to survive mankind's coming few centuries of climate change.

Of course, no one back in 1962 thought about greenhouse warming. In fact the 1960s were colder than average.

Some weeks later, I would see this same Royal Society Range from much lower and directly above. Professor Black and I were photographing, or maybe just observing, patterned ground in some of the valleys. We were flying in a Twin Otter, an aircraft, likely much improved and still used today. On our flight the motor suddenly stopped. We glided in terrifying silence. Did it stall or was the pilot just switching tanks? My panic was short-lived.

Our giant cargo transport landed on the McMurdo ice shelf; I think this large cargo airplane had skis. We climbed down the steps into minus-20 Fahrenheit cold. Off in the distance rose Mt. Erebus, an active stratovolcano at 12,448 feet elevation, totally glaciated in white, its summit crowned by a small white plume. This volcano was first sighted in 1841 by James Ross, who named it after one of his ships. A nearby extinct volcano he named Mt. Terror after his second vessel. The ice-covered sea on which we landed is called the Ross Sea.

We and our duffel bags were freighted to McMurdo Station in Canadian-built tracked vehicles called Nodwells.

McMurdo of 1962 was smaller than today, and less environmentally friendly. Back then, base power came from a small nuclear power plant, today a distant memory. Annual base waste was bulldozed onto the annual sea ice. We bunked in hemi-cylindrical Jamesway huts, comfortably warm at bunk levels but freezing cold on the floor. Antarctica back then was all male—mostly white males. No women and no cryo tourists. Local restaurant-wise, things haven't advanced. Even a McDonald's erected on a try-it, temporary basis couldn't make it, with the mega-harsh environment and absence of eager youth to dispense the burgers.

Of course the McMurdo we experienced in 1962 was a wealthy resort compared to the nearby huts built and used by the doomed Robert Scott (1901–'04 and 1910–'13) and Ernest Shackleton (1907–'09 and 1914–'16). These huts are, like McMurdo, all on Ross Island, at Hut Point, Cape Royds, and Cape Evans. When I was down there, some of these huts were accessible to visitors. Yes, even geologists can be tourists. I don't recall which hut I saw and entered, but it still had canned goods, a copy of Mary Baker Eddy and the chained half-mummy of a sled dog who presumably had been shot. As I recollect, one could only get to the huts by chopper.

In my first week or two the sun did set for a few hours, but it was never far below the horizon. Some of us drove a base Nodwell out onto the sea ice on an early spring day still below zero. We wanted to see the zone of collision between the moving ice shelf and the sea ice, which was piled high, maybe up to 20 feet, into jagged and jumbled miniature mountains. It was curiosity tourism, not research, on our own time. Weekends were not really observed, though we used to relax and goof off. The religious folks did have a small chapel for Sunday services.

These pressure ridges form where the moving shelf ice pushes into the sea ice near McMurdo station. Gaps with open water allow Wedell seals—lower right—to crawl out. Several of us drove there on the sea ice on time off.   Photo by Peter Vogt (late October, 1962).

I never shaved after arriving at McMurdo Station. Like most others, I grew a beard. Shaving required walking to another building. Besides, aren't Polar explorers expected to sport mustaches and beards decorated with frozen snot and breath condensate? I've kept a short beard for 60-plus years, save only briefly when beards became political symbols. I was no Marxist hippie—but have definitely become a lefty in my sunset years.

As November progressed, daylight grew to 24 hours. As the sun circled around us, dipping briefly only behind high mountains at midnight, we could tell time from which distant peak the sun appeared over or behind. There was rarely any light snow. Antarctica is a polar desert. Hard to tell new from old snow blowing around. The snowflakes were tiny. The air, usually in the upper 20s to low 30s Fahrenheit, was dry and the sky blue. Around Christmas time one could actually—out of the wind—take off shirts. Ross Island being volcanic, the dark lava outcrops warmed up. Of course on the high ice sheet itself, such as the South Pole Station, the mercury stays below zero even in summer. No, I never had the chance to visit the South Pole. Nor view the Geologists Range with its Vogt Peak I learned about later.

The field work was generally done by PhD student Tom Berg, with me as his assistant. Tragically, Tom Berg would die a year or two later in a helicopter crash. After the engine failed, the pilot brought it down by autorotation of the rotor blades, landing on Mount McLennan and sliding 700 feet down a slope. Six of the eight on board survived. The lightweight magnesium metal had caught fire, and Berg could not escape in time. Also killed was Jeremy Sykes, a New Zealander documentarian shooting a film. When the crash occurred, on November 19, 1969, Berg was beginning the final field season of a 10-year-long investigation of patterned-ground phenomena in Antarctica.

I don't recall major accidents down there during my stay. Driving a Nodwell to our research area on Ross Island one January day, I got stuck crossing a small frozen pond, evidenced by a photo earlier in this book. The constant sunshine had made the ice rotten, and I chewed its tracks down. I had to hike back to McMurdo to get help and have my Nodwell towed out by another.

On another occasion, camping and working on the other side of the Sound, I was carrying tools across a lake permanently frozen solid. I slipped and fell, my sledge-hammer landed on my right hand and broke a finger. The bone healed by itself and remains the only bone I've ever broken, *knock on wood.*

I knew of the poor Navy SeaBee chipping away the conical 'yellow glaciers' in the long outhouse who slipped and seriously broke a bone. He had to be airlifted back to the States.

Tell us again how you were injured in the line of Big Ice duty?

The head back then was a long outhouse, the seats not separated by partitions for privacy. I figure that changed after McMurdo became coed. There was always second-hand reading material; *Playboys* and an occasional *New Yorker*. Those lucky enough to poop next to a Naval Air person— maybe even a pilot—might hear some scoop on the next mail delivery from the States.

My other medical issue was an impacted wisdom tooth. Bad luck, but at least they had a dentist on station. The tooth had to come out, so I was sedated. However, I was not told beforehand that their big bomb-like Novocaine container had frozen during the last winter, making their Novocaine 90 percent ineffectual. It was far and away the most painful experience in my life. A 10 on the 1-to-10 scale. The Navy dentist kept trying to pull out a crooked root, and his pliers kept slipping, jamming the handles into the inside of my cheek. A muscular enlisted sailor held my arms behind my back. I kept blacking out. Then I heard someone say, "Are you having trouble with that tooth?"

Speaking was the dentist who had wintered over and had not yet left for the States. He took over and pulled it out with one big yank. By chance I met this skilled dentist's daughter at a Southern Maryland party more that 60 years later.

In his research on patterned ground, Black wanted to monitor the climate in his research areas. That meant installing a thickly insulated box with a number of truck batteries to keep the instruments recording through the bitter Antarctic winter. However he had to keep the batteries charged, and for that he was way ahead of his time. He had the bright idea of using wind power.

The generators powering Depression-era radios on the Great Plains were still being made. The company that made them, Wincharger, had made three-fourths of America's generators before the middle of the 20th century. So Black shipped several down to McMurdo, and I helped Tom Berg install them. The propellers whirred merrily in summer with 10-to-20 knot onshore winds.

All was well, and I checked on them before flying back to the States. Eventually, the Winchargers were blown to smithereens during hurricane-force katabatic winds draining super-cold dense winter air down off the Ice Sheet.

From Navy enlisted and grad students up to senior scientists, all asked themselves the same question: What Christmas presents can we possibly send home from Antarctica?

Why of course: stuffed penguins. Back then, they probably weren't made in China. They were about a foot high, the size of one of the myriad Adelie penguins communally chattering and pooping on their pebble nests in various summer rookeries. There were some big ones near the shores of Ross Island. In a major departure from reality, the stuffed penguins wore orange sashes across their white fronts, each one advertising itself as Souvenir of Antarctica. Their noses looked like stuffed carrots.

I had no legitimate reason to visit a penguin colony. However I managed to bum a helo ride with an ornithologist. Did I promise to help? Can't remember. But I took one of my two souvenir penguins along, minus its sash. When the penguin researcher was out of sight, I gently pushed an Adelie off his or her nest. I installed the pretender and went back to photograph what would happen. The angry nest owner waddled back and pushed away the imposter. And after the latter lay prostate beyond the nest, still pecking away at that obviously annoying orange bill.

Peter's stuffed intruder is evicted from nest by original occupant.

Today what I did would get me fined and sent back to the States. Yes, I was harassing the wildlife so shame on me. But I didn't hurt that penguin who lived long ago and likely remembered the adventure to his or her dying day. I gave away the mint-condition stuffed penguin to some cousin many years ago, but still have the imposter, decorated with poop dust a future owner could test for DNA to validate the story and ensure a higher price on eBay or Antique Road Show.

My other Penguin anecdote concerns a metal button proclaiming HATE PENGUINS that I took back home to Santa Barbara. My mother couldn't understand why that was funny and why anyone would hate penguins. Today, 60 years later, I don't understand either. Of course, I was not a penguin researcher. Maybe some humor changes with time.

The New Zealanders had their much smaller Scott Base within walking distance of McMurdo. One day I hiked there, not for any valid research reason but to mail a letter back to my parents. Why not simply mail it from McMurdo? Simple—in the latter case I would just use US stamps, but as a stamp collector, I wanted a rare postmarked New Zealand Antarctic Territory stamp. The US—and to their credit the USSR and modern Russian Federation—do not claim any part of Antarctica. Alas, that stamped envelope and my entire stamp collection would years later go up in smoke when our family's Santa Barbara house burned to the ground in the Sycamore Canyon brush fire of 1977.

The Kiwis at New Zealand's Scott base, Antactica, had rigged up a tow rope on snow as hard as rock. I took a day off in early 1963 and hiked there from McMurdo. To the right, steam wafts up from the Erebus summit (13,000 feet).

However, my Scott Base tourist visit did offer me more than stamps. The Kiwis had converted one slope into a short ski hill, outfitted with a rope tow. So I can claim to have once actually skied Antarctica. The fine snow had been hard-packed by wind into small drifts and my skis left no trace. It felt like skiing over highway speed bumps.

As I was the last of Black's small team to return to the States, I was left to crate the instruments and tools from the project. Minus, of course, those ill-fated Wincharger generators. I could still continue collecting some data, provided I found volunteers to help. I had already done so around New Years when I was entrusted with surveying, using a plane table, and some frost polygons in Taylor Valley. (That's also where I had earlier broken a finger.) Taylor Valley is one of a few Dry Valleys in the lower parts of the Royal Society Range across McMurdo Sound from the base. Those valleys had been glaciated at some earlier time, and one of the research problems was determining when that happened and why the glaciers are no longer there. Maybe modern climate change folks have found answers in recent years.

I had gotten to know some marine biology grad students, probably at the base cafeteria. My equivalents but not in geoscience. Their project involved chain-sawing square holes into the thick sea ice and lowering various instruments into the ocean below, maybe to the bottom, and collecting benthic specimens. The Antarctic (Southern) Ocean, unlike the land, brims with diverse living creatures! I tried to low-skill help them and witnessed their activities in their research huts. In turn they were curious about what we were doing and about that scenic mountain range across the sound.

Tent-camping in Taylor Valley

So I recruited two adventurous biologist volunteers as trainees. We were flown to Taylor Valley and dropped off by helicopter, to camp for several days. Rule No. 1 after touching down: Avoid decapitation by the helo blades. Rule No. 2: Hold on to camping gear and quickly carry it away, specially sleeping bags and foam rubber mattresses so those aren't blown away as the helo takes off for its flight back to McMurdo.

Camping in Taylor Valley was a great experience; sunny and dry, scenic mountains all around. No mosquitoes or other vermin, air temperatures in the 20s to low 30s. There was silence except for the wind, and not much of that blowing over our tents. Beneath was gravel, cobbles and sand, with nary a blade of grass or touch of moss. The cobbles—called ventifacts—had been sculpted on one side by sand-blasting from hurricane-force winter katabatic winds roaring down the valleys from the ice sheet.

Back then it was okay to pee in the outdoors and bury our poop, as we learned to do in Boy Scouts. But today, human waste is bagged and carried back to the base. The fragile, unique microbial ecosystems should not be further compromised. One night I had insomnia and went for a long hike alone. I found a sand-blasted crab-eater seal mummy. I suppose it got lost like that leopard in Hemingway's *Snows of Kilimanjaro*. There was one giant boulder left there by ancient valley glaciers. For some stupid reason I spray painted *Jesus Saves* on that boulder. That was 60-years ago, and I trust the wind has long ago erased my graffiti.

I won't forget New Year's Day 1963. We had shortwave radio and just for the fun of it strung out antennas on the ground to listen—if we could—to the Rose Bowl game in far-away Pasadena. I wasn't a football fan but in this game, my school, the University Of Wisconsin Badgers, were playing UCLA. We managed to pick up much of the game, and the Badgers won. Although my biologist volunteers were not Badgers, we toasted each other with cough syrup, the closest thing to alcohol we had.

Another souvenir of my summer on the Big Ice was sent to me at my high school address in sunny Santa Barbara in the late winter of 1963, when I had just begun my graduate studies in the University of Wisconsin in the Geology Department, near the frozen Lake Mendota. It was a letter from the Board of Geographic Names. I had authorized my dad to open my mail and forward whatever seemed important. He phoned to say he was very proud of his older son. He read me the letter, which said in part: "We have named a geographic feature after you—Vogt Peak" (+7,152 feet in the 34-mile long Geologists Range).

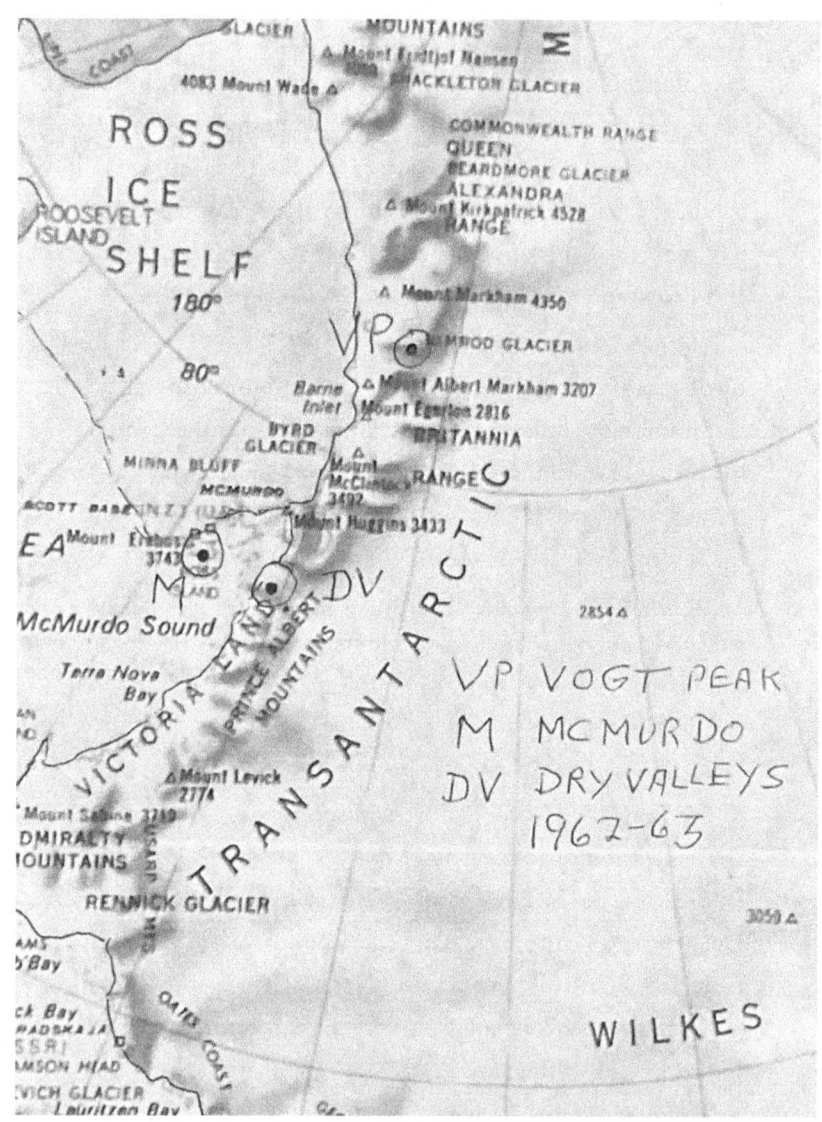

On this Antarctica map, Vogt Peak is shown on the right.
Vogt Peak (VP) is in the Geologists Range Range upper right on map.

Vogt Peak 82° 22'S, 156°44'E
Peak, 2,180 m. surrounding the E part of McKay Cliffs in the Geologists Range. Mapped by the USGS from tellurometer surveys and Navy air photos, 1960–62. Named by US-ACAN for Peter R. Vogt, USARP geologist at McMurdo Station, 1962–63.

These mountains had been seen by a New Zealand Geological Survey expedition the previous austral summer, 1961–'62. (The location of Vogt Peak is 82° 22'S, 156° 44'E.)

I had never heard of, nor seen, let alone researched this peak (a nunatak, almost all but the top buried under the edge of the great East Antarctic Ice Sheet). I tried to explain to my father that, with US Navy aerial photography, aided in places by US Geological Survey tellurometer surveys, so many new geographic features were being discovered that they had run out of important dead persons and even important living ones. The Board of Geographic Names evidently had to dig down into the bottom of the barrel and name mountains after beginning geology (actually geophysics and oceanography) grad students.

My dad thought I was just being uncharacteristically modest, so I stopped trying to take away his obvious pride. It turns out my dad also had something named after him: a small street in his South German hometown. Except he actually deserved this, having built a plane as a teenager not long after the Wright Brothers. His plane crashed after a few minutes airborne, not atypical in those pioneering times.

Letters Home

September 25, 1962, 10 a.m. PDT:
Postcard to parents en route
to Antarctica.
Sent from Travis AFB

It smells, feels, sounds, & looks (inside) like a Greyhound. Bumping & whining at 8,000 to 13,000 feet over the folded mountains of southern Wyoming & Salt Lake, variable cloudiness. Heavily laden, our DC-6, with Antarctic gear, 15 civilian scientists & servicemen, some of whom volunteered for the Antarctic and others who didn't. 12 hours down, 38 to go. Tonight we'll spend time at Hickam AFB (BOQ) Honolulu. After that only a few hours of fueling stops at Canton Is and Namdi (Fiji), then 'Chee Chee' (Christchurch).

CHAPTER 6

Back in the USA, Grad Studies Commence

We took off from the sea-ice runway off McMurdo in February 1962. It was about 25 degrees Fahrenheit in the late McMurdo summer. A long trip with several flights and tired jet lag brought me to Madison, where the thermometer shouted 10 degrees. A voyage from late southern summer to northern winter. I proudly wore the orange USARP (United States Antarctic Research Program) parka with its fur-trimmed hood.

Bascom Hall, the oldest building on the University of Wisconsin campus.

Even before looking for a place to live, I headed for the old red-bricked Science Hall to brief Professor Robert Black and deliver his most recent data. Only then would graduate studies finally begin.

Madison enjoys (or not) colder winters compared to Ohio, and borders that wonderfully expansive Lake Mendota, right next to the campus. That meant longer skating expanses until snow covered the ice. I witnessed speeding ice boats and ice fishers, but I never took up those sports. I had chosen the University of Wisconsin for several reasons: beer; a coed campus; cold icy winters: and, one more thing, the Polar Research Center. That connection enabled my Austral Summer: 1962 in Antarctica, then in 1965, plus two icebreaker expeditions to the west Siberian Arctic to collect data for my PhD thesis. (More about those cruises later.)

Before being covered in snow, Lake Mendota ice meant even more to me than outdoor ice skating. The lake ice was thicker than on that small Ohio lake I recalled from boyhood. Not only thicker, but even more decorated by cracks. Physics in action! Stress and strain. Tension and compression cracks were far more interesting. Tension created gaps into which water rose and froze. I later realized that ice tectonics is similar in many ways to plate tectonics: a miniature model. Tension fissures filled and frozen are analogues to the Mid-Oceanic Ridge, where plates move apart and molten magma rises and solidifies. However, plate tectonics had not been fully 'discovered' until 1968, the year after I left Wisconsin.

One winter day while walking across the lake, my future wife suddenly saw me drop down. Was her date having an early heart attack? No, I just wanted to look closely into the ice. She must have been having some doubts. Those doubts no doubt grew one subzero night after we were walking down Madison Street from a beer tavern, Glen-N-Anne's Cozy Inn, we frequented. The band, the Goose Island Ramblers, played polkas, with one member using a jug. Probably under the influence, I started to brag about my polar explorer status, and I suppose to prove my manhood I took off my shirt and walked bare-chested. I suppose my date wrote this behavior off to booze and harmless silliness.

Wisconsin did offer skiing, but no real mountains anywhere in the state. The local region ski hills were pathetic compared to Innsbruck or California or Colorado. One Wisconsin resort billed itself as Big Powderhorn. What bombast! Their ski hill was only slightly higher than a ski chalet. Okay, maybe I exaggerate. But I was inspired to invent and have made attractive ski patches that advertised Ski Big Bascom.

Bascom Hill was a gentle slope from Bascom Hall, the university headquarters, down to the nearest street. When it was snow covered, students would slide down the hill on cafeteria trays. My graduate student-shared office in the basement of Science Hall sported a window through which this spectator sport could be enjoyed. Some students, I recall, failed to bail from their trays in time and were launched over a four-foot wall onto the street. I suppose the sport has long been banned. However I do have some remaining unsold Ski Big Bascom patches for Badger skiers to wear in the Rockies or the Alps.

The best skiing in the Midwest is found on Michigan's Upper Peninsula in the Porcupine Mountains, known as the Porkies. They offer longer runs and more elevation drop. It is also very cold. With others from the University of Wisconsin, we would drive up there from Madison and ski in subzero weather. I recall that the tip of one ski cracked, maybe from the extreme cold. After skiing we went to a popular Finnish family's sauna and, as customary there, rolled warm in the cold snow upon emerging from the blistering sauna.

As young people—mostly guys—sometimes do, I took some truly stupid risks, which could have cost me my life. Was I looking to earn a Darwin Award? The lesser of two examples happened one warm spring day when I rode my bike out on Lake Mendota as it was thawing, overdue for its annual ice-out. The ice had turned to candle ice, and I figured it would not break if I rode along at a good clip. I rode along the shore in front of the University Union, where many students were sunning themselves and likely preparing to see me break through and drown. I'm not sure today why I did that; it was before I met my future wife, who would have concluded only that I was not her type. Maybe I did it just to show it could be done. The last skate of the year—on a bicycle. Might have been the last skate of my life.

Even more risky and stupid was my deciding to skate on the frozen Wisconsin River in the dead of winter. I was talked into this (or the other way around) by a tall athletic Outdoor Club member who hailed from a Baltic country. Estonia, I think. At that time it was part of the USSR, so he had likely escaped. This fellow—Andres Peekna was his name—sported a handle-bar mustache and, like me, liked outdoor ice skating. I was wearing the beard I had grown in the Antarctic. (Snot and breath condensate and freeze to beards in very cold weather, whether in Antarctica or Wisconsin.)

I'm sure snot froze to the tips of Peekna's mustache, but I had other things to think about. The Wisconsin River was indeed frozen, but not in the middle, where swift currents kept it open. Snow covered the nearshore ice, leaving only snow-free ice between open water and the snow-covered ice suitable for skating. However, the ice near open water was too thin. This left only a band of safe skatable ice along which we would skate. We parked our two cars some good distance apart along a lightly traveled road. I can't recall if we had ropes or spare dry clothes in the cars. We trudged through deep snow, carrying our skates, to reach the river banks. It was zero or below. We did the skate and obviously didn't break through and drown or freeze to death wearing wet clothes in subzero weather.

Would I ever again skate on a frozen river? Yes, but not one with flowing, open water, and not in subzero weather. And not really on a river. And with no snow anywhere in the area. And not in a place with cold winters.

CHAPTER 7

Amid Wondrous Beauty, Journey to an Ice-Capped Volcano

The Panagra 707 from Panama settles down into the neon sea of Los Angeles. A customs woman burrows through snake skins, fingers fake shrunken heads. And sniffs suspiciously into a plastic bag of coca leaves. At length the two North Americans, Peter and Volker Vogt, of Santa Barbara, staggering under their 20-kilo weight limit, climb into the waiting car of their parents and drive home.

A dusty, colorful summer lies behind them. After flying to Buenos Aires, where they spent several uneventful weeks with relatives dodging ancient cars and hoping for revolution, the two brothers set off on the long trek north...

For three days, Peter and Volker alternately pushed and piloted an overloaded Citroen Ugly Duckling over the mud roads of Argentina, arriving in the frontier town of Eldorado in time to witness a hysterectomy, performed and finished just before a power failure, in their cousin's clinic. They feasted on beef roasted over open fire, charred stumps of recently felled trees smoldering into the cool blue of a subtropical winter.

There was swimming in the deep, swirling waters of the Parana, where only a decade ago the deadly drone of malaria-bearing insects broke the stillness of the dusk. There was a night of Argentine drinking, dancing, and bombastic political debate on an ancient volcanic bluff overlooking a drought-weakened Parana, the lanterns of Guarani Indians twinkling in the lantana- and orchid-enchained forests of Paraguay across the river.

There, the liquid fire of Cana poured into one's brain to imprison reason and to liberate imagination. Background music of the Guarani cascaded into the sentiments echoing the myriad minor falls and cataracts that framed the boiling thunder of Iguazu, wilder and mightier than several Niagaras, barely 50 miles to the north.

One day, by the end of July, the two brothers climbed into an ancient and gaily painted bus and headed north again. Near the Iguazu Falls, an outboard ferry guided them through misty eddies to the Paraguayan side, Puerto Francom, from which daily buses made the six-hour ride to Asuncion. The capital was a charming, sleepy river town. Dirty tots crowded the cathedral steps to watch dictator Alfredo Stroessner's Indian youths goose-stepping in the afternoon sun.

The *Ciudad de Corrientes* glided over the Paraguay River's brown glass and throbbed downstream toward Posadas. Its course took us from the Argentine bank to the Paraguayan and back again. The scrub forest—southeast Texas a century ago— glided by, mile after empty mile.

Flamingos stalked in reedy swamps, and a squad of buzzards flapped heavily about some unseen carcass. An occasional log, disturbed by the wash of the wake, materialized as a crocodile and submerged into the muddy waters. Night came and we seemed suspended in a hazy, moonlit universe, while dark lines of shadowy trees floated past in an endless procession. The sharp clanging of a bell before sunrise announced our arrival at Corrientes, from where we hoped there were connections west across the Chaco.

The Argentine Chaco was virgin country for tourists. There were no Hiltons, no Ye Olde Native Crafts Shoppes, not even a *Casa de Cambio*. Between a sheaf of greenbacks and a 300-peso rail ticket, worth a piddling $4, lay an obstacle course that would blacken the day for the most intrepid traveler.

TO THE BOLIVIAN FRONTIER
ON THE GENERAL BELGRANO

It was morning; we travelled from the ticket office to the police to the stationmaster in vain. An endless variety of city people passed up the chance to profit 20 percent in the exchange. If the bill was really *un dolar* why did it say one, which looked like a misspelled *once*, Spanish for eleven.

Finally, after endless soliciting, haranguing and outright begging, we sold a few bills at a great loss. We were able to board the antediluvian wood-and-cast-iron coaches of *General Belgrano*, a rails antique named for Manuel Belgrano, the Argentinian politician who hastened reform in Argentina as the Age of Enlightenment swept Europe.

After a number of marrow-shuddering jolts, a mighty shriek awakened the morning mist and black clouds of wood smoke and steam filled the air. With a fiery snort the *General* lurched forward. Soon dust filled the nostrils and coated the lungs as we thundered across the Chaco at 15 miles per hour. The old *General*, after long, faithful, and profitable service at the hands of the British (greedy foreign exploiters), now lost money for his proud Argentine fatherland.

A six-hour wait in Roque Saenz Pena, a town the size of Ventura, California, gave us new opportunity to dump our cursed *norteamericano* currency at a loss. For those who contemplate running out of foreign currency, don't stray from the tourist herds. You might as well have Monopoly money as US bills.

Through a chat with the local bank *gerente* (manager), we discovered that provincial banks were not authorized to cash Travelers' Cheques. Of course they could be sent to bank headquarters in Buenos Aires, if we wanted to wait four weeks. But the man was sympathetic to the problem of impending starvation; after closing hours he shepherded us to a local Mormon mission where, he seemed confident, our compatriots would help us out. Four friendly Mormons, youths our own age, explained in a friendly manner that their church prohibited the loaning of money. Being Christian Mormons, however, they soon saw the light and cashed a $10 check with their private pesos.

For another two hours we parried their missionary zeal, then reboarded the Chaco Express. Through the dust, one could see *gauchos* with their broad-brimmed hats, flaring *bombachos* 'loose pants,' and accordion boots, conversing in a strange, cryptic Spanish at the other end of our car. Outside lay the endless brush-and-cactus plains and bird-studded swamps of the Chaco, whose steer skulls grinned at the tortured passengers, whose posterior pains were hardly comforted by beautiful cast-iron flowers in the baggage racks. Names of Chaco towns completed the setting: Monte Quemado, Rio Muerto, Pampa del Infierno.

Forty hours of travel lay behind as *General Belgrano* streaked among the Andean foothills at a fast walk. Soon the lights of old colonial Salta appeared. From the next seat a transistor-borne disc jockey at *Radio Odemes y Salta—Voz de la Tierra Gaucha* crowed over bargain panties, and we knew we were back in a civilized corner.

In Salta and Jujuy, gaucho *bombachos* mingled with the *bombines*, derby-like hats worn by the Indian women of the altiplano. It was still a day's trip to the Bolivian frontier. Slowly the train wound north through wild gorges, ever upward. The air became clear and thin and dry: Palms yielded to willows in the broad gravelly washes and great tree cactus watched from hilltops, like timeless Inca sentinels guarding the southern approaches to their empire. Far away, the icy crowns of four-mile-high Andean summits scanned the barren altiplano.

The frontier town of La Quiaca was a treeless collection of adobe boxes in the middle of a vast barren plain. Here, at two-and-a-half miles above sea level, vegetation consisted of a few grazed-off grass stubs, dead brown during the winter, or dry season. A tropical noonday sun stabbed through an azure sky, but in the shadows of the village well, frozen puddles survived the day.

Travelers should have had a Bolivian visa in their passport before they arrived in La Quiaca. These visas were issued gratis and without fuss at any major consulate. For some reason, we had neglected to get them. The morning after arrival, we puffed our way through the frosty air to the Argentine customs. There, a friendly fellow told us that before he could stamp a salida onto our passports, we had to exhibit the Bolivian visa.

So we tracked down a little stone building half covered by an enormous oval seal and the sign *Consulado de Bolivia, La Quiaca.* The church bell struck 10, at which hour the consulate was supposed to open, according to the sign on the door. An hour later the doors were still locked, and a handful of people had assembled. At length, a dirty-collared head appeared and announced that the Señor Consul was taking a bath; but that the doors would open shortly. At 11:15, 15 minutes before the Argentine customs were due to close for the day, we were still waiting.

Furious hammering on the door produced the same dirty-collared man. After some hesitation, he singled out the two North Americans, motioned us in, slamming the door in the faces of his countrymen. Passports were taken, and we shifted about uncomfortably. Next to the clock, which threatened the closing of the Argentine embassy, was a gaudy map of Bolivia, enlarged by the Littoral, a sizable piece of coastal desert. It, along with the ports of Antofgasta and Arica, were seized by the Chileans almost a century ago but still vociferously claimed by any patriotic Bolivian, who will tell you that since his country has a navy, it must have a port.

The grimy peasant shuffled in with the passports, still unstamped. It would cost $3 apiece. But the visas were gratis; we had a State Department leaflet on visas that says so. He demanded proof, and we couldn't find the leaflet. Protesting furiously, we were nevertheless at his mercy and so pulled out six dollar bills. He shook his head: It would have to be pesos, $6 worth at an outrageous exchange rate. We did not have any pesos, having changed them all to Bolivianos. He fingered the eighty thousand Bolivianos we offered him, thought for a minute, and then mumbled something about another dollar and a-half apiece because, it was, he said, *afuera del horario*— outside of normal hours.

At such times, the temptation to violence becomes difficult to resist. I grabbed the 80,000 Bolivianos out of his hand and, for the 'nth time, demanded to see the consul. *Come back in the afternoon*, he shrugs. The haranguing continued. Suddenly, the inner door opened and a small boy stepped out, motioning to our peasant. Little Fish scurries into the consul. Within moments he returned, telling us that $4 will do, took the Bolivian notes and began filling a passport page with an ugly visa, making a number of mistakes in the process. On the third line of his heavy handwriting, it read *Gratis*.

Brilliant moonlight illuminated the lunar fantasies of the altiplano, strange gorges, badlands, broad *salares* (salt flats), and snowy peaks. Villages were few. Llamas and vicunas roamed these grasslands, which climbed to the eternal snows; there were no pine forests between. The southern altiplano night was bitterly cold. The A 'road' repeatedly crossed the 14,000-foot contour. During the wee hours there was a half-hour delay. Ice in dangerous amounts had accumulated on dining car wheels. Steam jets removed the ice and we rattled on, only to collide with a mule some hours later.

The inside of our second-class car resembled a 19th century general store. Unwrapped soap balanced in high stacks on window sills. Lamb carcasses drooped from baggage racks, and restless chickens announced the coming dawn. We rode high on a stack of blankets, comfortably warm in several Argentine sweaters, smuggled goods loaned to us by the *chola* (Bolivian Indian woman), in the next seat, until the several customs inspections had passed. Goods of every sort made a trip to the toilet a major accomplishment. Description of the latter has been omitted, in deference to the spoiled readers of the *Santa Barbara News-Press*.

The train wound through the reed-dotted shallows of Lake Poope. Due to last year's heavy rains in the north, we rolled along cautiously for hours. Far away, glaciated sierras were visible along the horizon. Dust devils wandered forlornly along the dry plains. At dusk the still-sunlit form of 22,000-foot peaks illuminated our approach of La Paz. We were due to arrive in the capital in half an hour, but no lights were visible.

Suddenly, a fantastic sight appeared out of the blackness. The dark plain had vanished completely and in its place a thousand-foot void overlay the carpet of lights far below in La Paz. The train chugged cautiously 15 feet from the rim, as if it were hesitating for the awed passengers before leaping into this starry, unearthly sunken city. A careful check of brakes evidently revealed nothing, and amidst shouts in the frosty darkness, our descent began.

IN LAPAZ—BEAUTY, COMPANIONS

After two days and two nights on the train, we were tired, bewildered, and excited, at the station. Everyone seemed to know where to go, and soon the three of us—the Vogts plus a temporary traveling companion from Texas—were nearly deserted. After finding the nearest cheap hotel, appropriately called Hotel Ferrocarril, we wandered down into the holiday festivities on one of the plazas to get a snack. But bed was more attractive than fireworks, rude flute and trumpet bands, and chanting crowds, so we panted back up to our $1.75 room. The hotel manager had a different idea, however, informing us with a grin that "some girl friends of ours had called."

"Are you sure?" I said. "We know not a single soul in La Paz."

"Yes, they knew two of your names and that you were *norteamericanos* who just arrived."

Needless to say, curiosity won over fatigue. The ensuing telephone conversation revealed an apparent and improbable mix-up on which we capitalized in arranging to meet the girls the next day. The meeting proved to be a fine intro- duction into the upper middle-class social circles of La Paz. Our gratefulness at this opportunity was surpassed only by the eagerness with which our various hosts competed for the three *estudiantes norteamericanos* during the next 10 days.

Lunch, tea, dinner, and evening invitations followed in rapid fire. Included in the profits was an overbearing maternal care that descended on the one of us unfortunate enough to fall ill. The fact that there was a medicine and biochemistry major among us did not tone down the insistence that a particular elixir would cure the unfortunate.

Besides learning Bolivian dances and the American Twist, we found intellectual stimulation in listening to a Bolivian Navy officer talk of the virtues of national socialism and of the need for a dictator in underdeveloped Bolivia. (Indeed swastikas far outnumbered hammers and sickles in the whitewashed walls of La Paz.) Other events in La Paz included a day's excursion to Chacaltaya, the world's highest ski resort at 17,300 feet, where we were delighted to find a crew of American scientists working on cosmic ray showers.

The Independence Day celebration lived up to the expectations of Latin American patriotism and fervor. Speeches reaffirming Bolivia's right to the Littoral were followed by goose-stepping units of the armed forces, low-diving Mustang fighters, and a lone black widow bomber that skimmed over the crowd amid approving roars.

Eager once again to be free, we severed the newly formed bonds of mixed friendship with promises to write and climbed aboard the train to start the 36-hour journey to Cusco, Peru.

HIGH IN THE WORLD OF THE INCAS

The night steamer across Lake Titicaca, altitude 12,000-feet, having no more second-class berths available, we snoozed successfully in the first-class lounge.

The difference between Bolivia, which had its revolution in 1952 and had redistributed much land, and Peru, where still the 40 rich families controlled much of the wealth, stood out clearly to the perceptive traveler. Rude huts scattered over the land gave way at the border to groups of large, uniform fields nucleated by a hacienda and many small houses. That we came in contact with no great discontent did not belie the fact that Peru remains a powder keg, despite reform-minded President Belaúnde Terry.

Cusco was a charming and amazing city, not unlike Santa Barbara in its environment. Greener and more mountainous than the barren south, the altiplano here appeared less inhospitable. Unfortunately the old Inca capital was a little overridden with tourists (at least as South American cities go), as evidenced by the high fixed prices.

Spurning cabs and guided by an Indian boy with an amazing knowledge of his Inca ancestors, we trekked along a 10-mile circuit to visit the four chief ruins outside Cusco. Descriptions of the functions of the various parts of the ruins were spiced by eye-witness accounts of the ghosts that still came wandering through the 200-ton foundation blocks that the Incas somehow moved into place. There, as well as in the Inca walls upon which Cusco itself was built, it was impossible to insert even a pin in the mortarless crack between the closely fitted stones.

A trip to Cusco could not end without a visit to the Indian market held in Pisac. We decided to go as the Indians go—by truck. But even after a couple of thousand miles on overcrowded second-class South American trains and buses, we were not prepared for what followed: A full 60 men and women packed themselves so close that shifting weight from one foot to another proved to be a formidable effort, not to mention stepping somewhere else.

The color of the Pisac market was enhanced by the bright sport shirts of the many foreign tourists who came by cab to snap pictures and bid up prices. Here, even the usual South American password *pobres estudiantes* did not soften merchants' ears and lower prices below the high level in Cusco.

Having secured the last three bus seats available for the trip to Lima on Tuesday, we were left with only Monday to take in the world-famous lost Inca city, Machu Picchu. To be sure to get tickets on that train, we left money and thanks with an American girl to buy them on Sunday. But she failed to obtain tickets and was left only with the promise of the ticket clerk that there would be tickets left Monday morning. We were dismayed to learn Monday that all tickets had been bought up by the tourist agencies, always given first choice.

Pleading with the station master produced no results, and so we decided to gamble by boarding the train in the rush. Once underway, they certainly wouldn't stop the train to kick us off. Or would they? We hastened to make some quick friends, both American and Peruvian, in case an argument ensued. But beating hearts were soothed by the calm conductor with special tickets ready to sell for 501 pesos more than the already tourist-oriented price to a few unfortunates like us.

As the train wound its way down the Urubamba toward the eastern jungles under towering 22,000-feet glacier-locked masses rising out of nowhere, pangs for wasted money disappeared. Accentuated by the mists and smoke of burning jungle, the unbelievably steep mountains became still more precipitous. Like a Japanese painting, cliffs climbed out of unknown depths and reached with their icy fingers for the sky.

The lost city, discovered only 50 years ago, is perched on a 15,000-foot ridge encircled three-fourths by the snake-like Urubama below. It may never be known why the Incas chose this inaccessible spot for their city. Rising another 1,000 feet above the citadel is Huayna Picchu, a peak so steep that the Incas cut steps—called the stairs of death—into the rock to lead up to a fortification at the top. Outstripping the camera-clicking of the guided tour, the Vogts, both hiking enthusiasts, raced up and back down just in time to catch the return train.

An unforgettable bus ride from Cusco to Lima began the next day with the three *norteamericanos* occupying the hindmost three seats—on a long lever arm behind the wheels—in the 1963 Ford bus. Only new buses were used on this trip, explained the natives, because of the grueling nature of the road. Up over passes and down into gorges and up again climbed our vehicle, following a dirt road that became almost indistinguishable from the barren heights around. Fantastic vistas and freezing cold mixed with illness and weariness to sear these pictures into the mind.

A live goat tied to the top of the bus fell off and was retrieved unhurt. In a freezing dawn at 14,000 feet, passengers warmed themselves with tea while hoar-frosted llamas waited patiently for the burning sun. The bus careened around corners fenced only with wooden crosses where two buses with all aboard had been lost by the same transport company. Finally, after averaging seven km/hr for 36 hours as the crow flies, we reached the coast and asphalt of larger towns, and hence raced on to Lima, arriving in the misty dawn after 48 hours of bouncing on narrow seats.

In Lima, as in La Paz and Cusco, we lived in a cheap, second-class hotel and used the centrally located ritzy tourist hotels as a base of operations (supply house and comfort station). After visiting some of the world's worst slums, collecting shells at Callao and fishing a gift collection of stamps out of the trash into which the hotel boy had mistakenly thrown them, we were off again to the north along the coastal stretch that made the Mojave desert look like a garden of verdure.

The approach of the Ecuadorian border a hundred miles away was signaled by checking of passports, again, and again, and again. Relations between the two countries had been strained ever since Peru forcibly seized an oil-rich section of land from Ecuador some 30 years ago. After a full 20 checks and six stamps, we boarded the night boat to steam across the bay to Guayaquil. Snoozing in the comfortable boat hammocks till daylight, we disembarked finally at the rat-infested harbor, where anything in the world may be bought.

From tropical Guayaquil a bus worked up to the altiplano again. Here to the north the plateau was somewhat lower and much greener. Like the last Inca, Atahualpa, we found it more beautiful than the south. As the bus ascended from the tropics through thousands of banana groves to the dry wind-ridden plain and finally crept over the pass, the staggering sight of the highest active volcano in the world confronted the travelers. Cotopaxi at 19,000 feet had erupted as recently as 1917. We came, and saw, and decided then and there to try to conquer.

With the aid of a school friend an expedition came hastily into being. The laden US Embassy jeep clawed its way up to the 16,000-foot contour, finally gasping to a halt. We began gasping at that point, with nearly half of the atmosphere beneath and heavy packs above. The night passed unwillingly as the puny men shivered in their tents at the permanent snow line with the gale-winds forcing fine volcanic dust and cold into every niche.

I could not sleep; I was just not getting enough air whenever I rolled to one side. A wind filled the thin air with fine dust.

At 2 a.m. the word was up, and so the cold, unsuccessful trek began. I walked a few hundred yards, nauseated, and asked myself why the hell I was doing this. We humans should challenge ourselves now and then. I get that. But there were limits. I decided on challenges—and limits. I went back to the tent to sleep and wait for the others.

They returned without having reached the summit. Poor conditioning, illness, and newly formed and uncharted crevasses turned back the group at near 18,000 feet, just below La Llanasacha, the towering black cliff just below the summit.

We were glad to have tried, however, and were comforted by the frequent failures of others.

Once back in Quito again, we had a little time to explore that ancient city. Like so many South American cities, it was a harsh blend of the new and the old, the rich and the poor, the beautiful and the ugly. New California ranch-style homes homes and cheap tenements alike were cut off every afternoon from the dwindling city water supply. Somewhere, nobody knows where, perhaps underneath the most exquisite churches in South America, flowed the city sewage in a 200-year-old wooden system whose plan no longer existed.

As the last days sprinted toward the end of the summer and the return flight, a final hike up Pichincha overlooking Quito became reality.

Twenty hours after enjoying the panorama at 16,000 feet we were in the hot thick air of Panama, reminiscing about the adventures past and planning for a future trip.

PART II

The Sea: Exploring the Depths

Humans took to the seas long ago and for many reasons. Until the 1800s, however, going to sea for pleasure or science was scarcely a motive. This book recounts the experiences of an ocean-going geologist during the second half of the 20th century. The Age of Exploration was history—but that was just geographic exploration. The greatest paradigm shift of science—Charles Darwin's works of 1859 and 1871—were also history. However, when I first studied geology in the late 1950s, the great majority of geologists did not believe continents had moved around.

I not only experienced but participated in the second greatest paradigm shift in mankind's understanding of Earth—the recognition of plate tectonics, in the late 1960s. The strongest evidence came from the ocean floors. Also, the US Naval Oceanographic Office had a treasure trove of relevant data. It was my good fortune to be offered a job there, and I have few regrets not entering—as most fellow students did—the academic world. I came to the Navy civil service in 1967, during the Cold War, and Anti-Submarine Warfare was and long remained a major context. Twenty-two years later, the Berlin Wall came down on the precise day, November 9, 1989, on which I disembarked on a Norwegian research ship.

The world and ocean-floor research technology have changed in any number of ways. Computing power and autonomous underwater vehicles stand out. Deep sea drilling for science was just a plan in 1967. And when I began, US ocean science was still dominated by white males (and sometimes women).

For ages, many kinds of seamen (and women) went to sea. There were those who manned the sails and oars, and later the boilers and engine rooms. Merchants and naval warfighters; whalers and fishermen; emigrants, adventurers, and slaves —all went to sea. So did the many seeking new trade routes and riches. Untold thousands of them died at sea—of scurvy or warfare, or in the most ignominious fashion of being eaten by cannibals. So many descended to Davy Jones' Locker when their ships sank in storms.

Seamen might come back after many years. Those who returned often told sea stories—including tall tales of mermaids and sea monsters, and stories of strange lands, creatures, and peoples.

Going to sea to learn about ocean currents or seafloor geology only took off in the middle 1800s. There were some antecedents. Classical Mediterranean sailors heading for Egypt already used water depth and bottom type to navigate the shallow waters beyond the Nile Delta to Egyptian ports. Christopher Columbus, in 1492, already knew about variation of the compass: the fact that the compass does not point to the geographic North Pole. Ben Franklin used logbooks of ships to discover the Gulf Stream.

The depths of the open oceans were long unknown. On Magellan's early 16th century global circumnavigation, a

500-fathom check-line was fully lowered at one perhaps arbitrary point in the western Pacific. Although the line did not reach bottom, the point was recorded as the "deepest point in the ocean." Mariners then and now have been mainly interested in water depth being sufficient to avoid running aground. Especially in estuaries like Chesapeake Bay, research vessels can't get close enough to shore. Pilots and skippers may lose licenses if they run their ships aground. Thankfully, sediments flooring the Chesapeake Bay, along which I live and where I wrote this book, are mostly soft sands and clays, without big rocks.

CHAPTER 1

First Voyage: Dishwater Coffee and Greasy Food

In the summer of 1958, when I was 19, I volunteered for a Scripps Institution of Oceanography student fellowship. No pay, bad greasy food, coffee like dishwater, during 45 days at sea with no port stops. This was the Research Vessel *Horizon*, built for U.S. Navy auxiliary use in 1944 and serving Scripps scientists from 1949 until 1968.

At 143 feet long with a 33-foot beam, RV *Horizon* was small for eastern Pacific seas and swells. Compared to the 1,188-foot-long cruise ship *Wonder of the Seas*, *Horizon* wouldn't even stretch across its 210-foot beam.

Aboard the *Horizon*, work proceeded around the clock.

Our chief scientist was Dr. John Knauss, a physical ocean-ographer. Our mission was to track the deep, mysterious Equatorial Countercurrent by dropping Swallow pingers, aluminum cylinders with batteries inside, invented by English scientist John Swallow. They are weighted to float vertically and sink only to a certain depth. The pinger (transducer, like on common fish finders on Chesapeake boats) is suspended from the cylinder, and pings at regular intervals. The *Horizon*'s transducers would pick up the sound and track the movement of the deep pingers as they were carried along by the current, much as a weather balloon is carried along.

The trick to getting pingers to sink only to the depth of the current but not deeper or shallower involves careful weight adjustment. Making the pingers neutrally buoyant at Pacific countercurrent depth was my job before we sailed. I dropped just the right number of lead shot pellets into the pinger as it floated in a tank filled with water with the density (based on temperature and salinity) of the countercurrent.

At sea, work is vastly different from on land. True, the commute from bunk to lab or deck is very short. However, there are no holidays, no weekends, no TGIF days.

Work at sea goes on round the clock, for the officers, the crew, and the scientists. Everyone (except the skipper and cooks) stands regular watches. Typical is four hours on and eight hours off, but sometimes it's 12 on and 12 off. In my student days I preferred the graveyard shift—midnight to 4 a.m. Not too many people, especially supervisors, in the rather cramped lab.

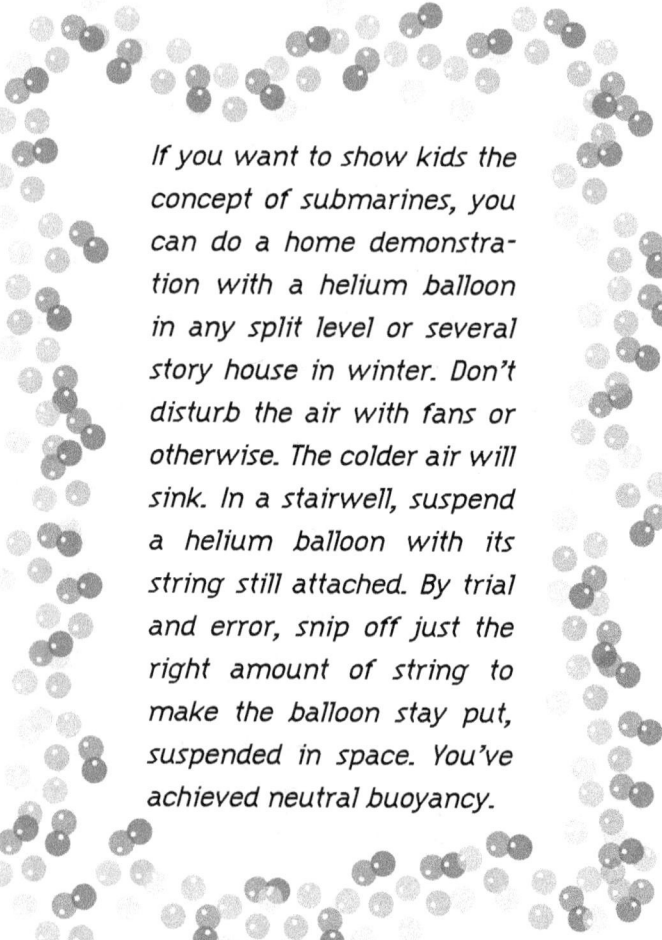

If you want to show kids the concept of submarines, you can do a home demonstration with a helium balloon in any split level or several story house in winter. Don't disturb the air with fans or otherwise. The colder air will sink. In a stairwell, suspend a helium balloon with its string still attached. By trial and error, snip off just the right amount of string to make the balloon stay put, suspended in space. You've achieved neutral buoyancy.

At night on the afterdeck, the ship's wake might be lit up with bioluminescence. In tropical waters I often tossed back overboard flying fish that had landed on the deck, which was low to simplify launching and retrieving cores, water sampling bottles, rock dredges, and towed magnetometer sensing heads.

As well as standing watches and performing experiments, the science party secured the instruments on leaving port before the vessel got out on that rolling sea and prepared them for operations once clear of the harbor. Launching overboard and retrieving gear on a moving research ship was no joyride, and if not done right, dangerous. This work on deck required coordination with the winch or A-frame operators.

Yes, I was seasick but did learn to swear like a seaman, lay figure-eight cables on the deck (works well with garden hoses so they don't get tangled) and stay way clear especially of taut wire cables. Should one snap, it could cut a person in half.

Back home in Santa Barbara, my mom kicked me out of the house until I cleaned up my seaman language.

I swore never to pursue an ocean science career and did not go to sea again for the next six years. Time, however, tends to cure unpleasant (but not traumatic) memories.

CHAPTER 2

Fate Joins Me
With RV Chain

RV *Chain*, a converted Navy salvage tug docked at Woods Hole
Oceanographic Institution on Cape Code, could hold 65 scientists
in addition to crew.

During the Wisconsin fall semester 1963, I was still undecided
between glaciology and land geophysics. My unpleasant 1958
ocean experience aboard RV *Horizon* was almost forgotten.
That expedition, about currents, had nothing to do with the
geology below, where my interest lay.

Fellow grad student Francis Birch, who had contacts with
Woods Hole, planned to sign on as research watch stander on
one of their expeditions. WHOI (pronounced *Hughy*), he
mentioned, was looking for a student watch stander on its 1964
marine geophysical expedition to the Indian Ocean, one of its
contributions to the International Indian Ocean Expedition.

The idea of an international expedition to the Indian Ocean had been hatched at WHOI during the August 28–30, 1957, meeting of the Scientific Committee for Oceanic Research inspired by the 1957–'58 International Geophysical Year. The Indian Ocean was chosen in part because it was the least known of the world's three major oceans. Recognized as the "greatest multinational effort in oceanographic history," the expedition involved 40 vessels from 13 nations comprising almost all disciplines of marine sciences.

My fate joined me with the *Chain* 43 expedition focused on marine geology and geophysics in the Indian Ocean, with research conducted en route via the Central North Atlantic, the Mediterranean and Red Seas, and returning the same way.

RV *Chain* was a converted Navy salvage tug, 213 feet long and 41 feet in beam. She could handle a crew of 29 in addition to 65 scientists, though we never had that many onboard. *Chain* would serve Woods Hole from 1958 until 1975. Our planned six month expedition would be the longest deployment of that research vessel up to then.

Hours were spent in the lab where information was processed
on RV *Chain*.

She was outfitted as a floating research laboratory, carrying nearly every marine geological and geophysical tool available in greater variety than on modern research cruises. Underway surveying, a gravimeter would measure the force of gravity, sea state permitting. The magnetic field strength would be measured via a sensor (called a fish) towed far enough (ca 750 to 1,000 feet) behind the vessel to get away from the ship's own magnetic field. A velocimeter would record the sound speed, and a towed thermistor chain the sea water temperature.

A sparker towed behind the ship would discharge an electric charge from onboard capacitor banks into the ocean every 10 seconds, causing sound waves that would penetrate not only the ocean but also through soft sediments below the seafloor and thereby measure their thickness. Those bottom and sub-bottom echoes would be picked up by a towed hydrophone array and graphically displayed in the main laboratory on a Precision Graphic Recorder and stored on tapes.

Chain would slow and hover at stations over selected ocean floor spots to lower to the seafloor devices attached to long steel cables. Cameras would photograph the ocean bottom, dredges scrape up rock samples, and tubular core barrels penetrate and recover bottom sediments and measure heat flow coming out of sediment-covered seafloors by attaching thermometers to some core pipes. Launching or retrieving a camera, core, or dredge— especially in a rolling sea on a wet deck—was a delicate and potentially dangerous operation.

Sometimes the ship's inflatable Zodiac would be launched to deploy a sonobuoy some distance from the vessel. The ship would power up its sparker, and the sound waves would pass into and out of bottom sediments to be recorded by the sonobuoy and radioed back to the ship for analysis. The result would be the sound speed structure of the seafloor sediments. The Zodiac would then go out and retrieve the sonobuoy.

The *Chain* 43 science group—typical for that time—comprised a Woods Hole Institute chief scientist (on some legs two co-chief scientists), about 10 watch-standers, a computer specialist (Dave Powers, an IBM employee), a technician (Barry Martin, employed by Woods Hole), and some scientists/PhD students who came to conduct special research projects.

While seismic reflection profiling in the Mediterranean we watchstanders noted sub-bottom dome-like structures in the sediment column below the ship. Maybe salt domes. The recordings were vertically exaggerated and the features looked indecent. We didn't know what to call them in our watch-stander logs, so we invented the term 'sullaf.'

Our chief scientist at the time was the venerable John Brackett Hersey of Woods Hole, very conservative and a deacon. He asked about 'sullafs' and we told him it was probably some Arabic word. It took some time before he realized it was 'phallus' spelled backwards. Dr. Hersey was not amused and ordered watchstanders to delete our neologism.

Watchstanders were mostly graduate students from various universities, and not from Woods Hole, which was not a degree-granting institution. There were two snappers who supervised the watchstanders. Snappers could be a co-chief scientist or a more experienced watchstander. Many, including chief scientists, guest scientists, some watchstanders and those with special experiments, were only onboard for only one or several one-month legs. In our age of air travel, it was no problem to fly to and from the various ports. As watchstander, I would serve for the entire six months.

I was primarily responsible for the magnetic measurements, but I was also part of a team working under a chief scientist, standing watch over other instruments and helping with station work and plotting data.

According to plans, the vessel would stop at Ceuta (Morocco), La Spezia (Italy), then via the Suez Canal to Aden (Yemen), and on to Mahe (Seychelles), and finally, Mauritius. Then back to the Seychelles, and via the Red Sea and Suez Canal to Beirut (Lebanon), La Spezia, Monaco, Plymouth (England), and back across the North Atlantic.

In Southern Yemen with friends.

I was single, and the many exotic port stops, about a month apart, beckoned. Maybe an avenue to a master's thesis! I applied and was duly hired.

I traveled from Madison to Woods Hole by bus, spending a night first in a late movie and after that in a seedy bus station. I arrived almost two weeks prior to departure, so there was plenty of time to check out the Captain Kidd Bar.

RV *Chain* backed off the dock at 4:15 p.m. on February 15, 1964. It would not arrive back there again until eight in the morning on August 21.

PASSAGE I:
FROM MISERY TO MOROCCO

Leaving the Woods Hole dock, *Chain* headed into a winter North Atlantic storm. With 20-foot seas and the boat rolling, there was a fire in the lower lab due to a sea-water short circuit. Seasickness was common. I spent the first full day in the rack. It was probably the most miserable time on the entire cruise. Why was I out here? Would I have to endure a half year of this? The malicious watch supervisor ordered me out of the rack to stand my watch.

Even a relatively small research ship is a labyrinth. I commuted from our cabin (the science team bunked at two or four per room, with a toilet and sink for each cabin) to the main lab, visited the mess for meals, helped in station work on the fantail and sometimes went into the pitching bow to the washing machines. If I visited the engine room out of curiosity, it was not long—it was usually hellishly hot. I went up to the bridge sometimes, when custom allowed. It was interesting for us scientists to chat with the officers or skipper, enjoy the view and peer into the radar screen.

Three meals were served in the two daily messes at regular local time hours. The officers and scientists ate in one mess and the crew in another. We watchstanders ate at one end of the officers' mess. The ship's officers ate in the middle, while the skipper sat with the one chief or both co-chief scientists, at the other end of the officers' mess. Tables were covered by rubber mats and other measures to keep the food from sliding off the tables. However during really big rolls (our biggest was 35 degrees), diners and table food could be plastered against the nearby bulkhead.

The food was excellent, a far cry from the *Horizon*. The cooks had tyrannical power and had to be humored. If you incurred their wrath, you'd had it. Early on, I was standing on the fantail next to the young Mississippi messman. I remarked that the waves were high but not as high as yesterday. In pure Mississippi dialect, he looked derisively at me and remarked: "Them ain't waves, them is seas." My physics background had gotten the better of me.

On February 28, 13 days at sea, the shadowy bulk of Madeira appeared a few degrees off starboard. By midmorning, the decks were peopled with curious, land-starved seamen, armed with binoculars and cameras. The seas had incised spectacularly into thousand-foot cliffs in the red-brown volcanic rocks of this ancient shield volcano. Vineyard terraces and small white-washed villages appeared, plastered on the steep hillsides, each one marked by a church steeple.

It was a sparkling, brisk day with a few low cumulus clouds fringing Madeira's summits. We rolled all day in mountainous swells, Madeira passing into the sunset. Late evening a jagged junior partner, reminiscent of serrated volcanic mountains in southern Arizona, somberly sailed by on starboard. Flying fish and porpoises rode the bow waves during the day. The smell of land reached us. At night, the full moon's highway of silver was hyphenated by cloud shadows.

On March 2, in light rain at 9 a.m., we steamed slowly into Ceuta Harbor in Morocco with a pilot aboard. Though we waited endlessly, the official okay to disembark never seemed to come. Land-starved sailors climbed off nonetheless, the first being the ship's crew, who promptly headed in the wrong direction. We set foot on a wet, greasy concrete dock. Oh joy: unyielding terra firma!

Over a day of sightseeing, a brilliant African sun and bits of blue sky replaced the cloud deck. Great billowing cumulus clouds moved above, but showers continued and rainbows appeared. For lunch at the outdoor tavern, I had *filet a la plancha* and French fries, with a fish appetizer and Africa Star Beer. My shipmates would obtain much more alcohol—cases of duty-free Scotch and Heineken for the next legs of *Chain's* voyage.

PASSAGE 2:
STRANDED IN ITALY
(OH THE MISFORTUNE)

A three-day stop in La Spezia, Italy became longer, causing us to tarry from March 10 to 14, when the radar orientation system developed a flaw and we had to wait for parts to arrive from the States. Most of us on board enjoyed the bonus shore leave exploring Italy.

Saturday about noon, the *Chain* moved from the Italian Navy pier to one of the commercial piers, the last boat in a row of fishing trawlers and tugs. Collections of strollers walked by to look her over; perhaps the news media alerted them to this rare vessel in their midst. Balmy sunshine flooded the glistening new grass of spring, towering cumulus hid the snowy marble crags of Carrara.

That afternoon, docking technicalities over, the scientific party dispersed to Florence or Rome, and myself to Siena, replete in Medieval and Renaissance splendor. There I brooded on a hilltop lorded over by the city hall tower and the striped Romanesque cathedral, its candy contrast heightened by a chill rain. The train, which went over Pisa and changed in Empoli, cost 1650 lira.

The olive trees and grass-green fields wandered by to the timeless, regular clacking of rails, or galloped into the past to the urgent, ear-splitting whine of the double-D express from Rome to Genoa. Fortressed hilltop towns framed in cedars and Respighian pines glided by, as if a moving stream of Renaissance landscapes.

Chianti bottle in one hand, chunks of bread and Formaggio in the other, I dreamed through the Italian countryside, when I heard, *WHAT?... I am addressed in English.* An American ceramics engineer stepped forward from the passenger milieu, a discussion ensued, and I spent the evening in Pisa with an American family in an apartment overlooking the Leaning Tower.

He and his bubbly extroverted wife enjoyed Italy and counted on making fortunes, buying a boat and sailing into relaxation. We talked of the sea, and both admonished me not to marry for quite a while. I might, no I wouldn't, no I don't think so, no, no, NO! More and more I count the blessings of freedom.

On Friday the 13th, our radar part arrived—in Rome, at midnight, further delaying our departure. I decided to visit Parma and for some reason, my diary pages were written in German, which I learned back then. Sixty years later, I must translate back to English.

It reads:

> I wandered downtown and found the hotel
> Terminus—a clean one-bedroom with shower,
> for 1,750 lira. I had dinner with wine, formaggio
> cheese and a mandarin for dessert, as normal.
> Then a cup of chocolate caldo in the espresso bar.
> Then I went to sleep under a blanket.

The next morning. I bought two thermometers for the ship's lower lab and visited the Renaissance Gallery. I wandered among some unremarkable churches and climbed back into the train car. Under the influence of red wine, the sounds of the train seemed to hypnotize.

Finally, on Saturday night, a huge crane pulled up alongside *Chain* to lift the radar part in place, replacing the antenna.

Underway, finally, we steamed down the Golfo de la Spezia, calm and sunny weather turning to heavy seas, the first rays of sun reddened the leaden cloud banks. We passed within eight miles of Stromboli, low overcast, nothing seen. Only in the shadow of towering rugged mountains did the ship turn into the Strait of Messina which, at its narrowest point, is barely two miles wide. One wonders how many wrecks sleep here since classical times. Getting a sailboat through, past Scylla and Charybdis, seems no mean task. Steady westerlies gave way to erratic gusts alternating with dead calms. Here the sea's face was oily smooth but swirling with eddies of currents; a hundred yards away turbulence and whitecaps.

The gray hills of Sicily were brushed with green, and a heavy aroma of spring and of land was in the air. The coastal villages, treeless collections of white stucco boxes, seemed unreal. Etna, in the distance, was socked in clouds, although her snowy flanks suggested a noble summit.

Sicily, Italia, and clouds soon disappeared astern. The sea was still disturbed. There were rumors of gale-force winds tonight, and loose articles were secured. But the storm failed to materialize, and morning arrived with a pastel, almost tropical dawn.

PASSAGE 3:
AT THE SUEZ CANAL—
STRANGE SPECTACLE, MORGANA EFFECT

On the evening of March 19 we rocked gently at anchor in 50 fathoms of water, awaiting clearance to pass through the Suez Canal.

Probably 50 ships, mostly freighters and empty tankers bound for the Persian Gulf, lay at anchor over the muddy waters around us, an imposing glitter of lights. Astern several hundred yards, the smart maroon tanker *Jetta Dan* flew the Danish flag. Toward port, the *Orient Clipper,* home port Malmoe, with blond, blue-eyed Viking lasses carefully observed through binoculars.

Abeam rode the *Moniuszko,* flying the red-white Polish eagle. We watched their officers on the bridge, their binoculars fixed on this curious American tug. The light blue Scandinavian *Merck Marie,* whose drunken seamen we had met in La Spezia not long ago, was also in the waiting fleet. One Norwegian seemed loaded with yellow school buses. The only other American and most decrepit of the bunch was a rusty old truck-carrying freighter, *Alcoa Voyager. Chain* was THE unique vessel, no doubt. All the ships flew the yellow quarantine flag, as well as a black–yellow striped one indicating 'waiting for a pilot.' The *Chain* also flew Stars and Stripes aft, UAR (United Arab Republic) forward and a brightly colored line of semaphore flags giving our call letters.

The fellows from science, having tired of jumping rope on the fantail, gathered in the scientists' day room to strum guitars, read and drink duty-free Heineken from Ceuta, Italian beers, and chianti. Not to mention orange soda, root beer, and ginger ale, of which the *Chain* seemingly had an endless supply aboard.

The advantage of being in the scientific party was really apparent at such times: Nothing to be done. The ship's people work their eight-hour day regardless and get overtime pay for every extra minute. If nothing important needs doing, they scrape rust and repaint, a job that never seems to be completed.

The chief scientist was getting his locks clipped by the second cook amid the songfest. Some of the guys stretched out on their flimsy Italian hammocks. Sprawled on top of the hawsers on the top lab, I watched the canalscape glide by: roaring Cadillacs; Egyptians on their asses (riding or just plain sitting); and an oxen plodding around a well shaft, pumping fresh water from the marshy lake west of the canal near Port Said to the tiny, irrigated cotton and grain plots along the western edge.

Also: a clump of palms; a row of gray-green, mesquite-like desert trees; a grass-and-mud brick hovel and barn; and a noisy diesel train, with a load of waving kids riding on top of the engine. To the east, a saltwater marsh and tidal flat, with white gypsum beds and sand dunes shimmering in mirages far away. The air was still cool, but a bright sun played hide-and-seek with cumulus clouds. A bunch of brown kids in

Arab 'pajamas' grinned and danced on the canal side street, thrusting their genitals at the American ship as a gesture of respect and goodwill.

A number of Egyptians were aboard: the pilot; a young research officer interested in pursuing studies at Woods Hole, the one for whom the depth profile was being made; another official; two oarsmen; and a Nasser-like electrician, a man whom I traded, for 250 Italian lire, a plastic comb, a roll of cellophane tape and a bar of Dial soap. I also gave him four pencils to mail my letter home, something he probably won't do. The fate of this correspondence is unknown: if the Arab mailed it, the letter would have been burned when my parents' house burned down in 1977.

The two oarsmen served only, as far as I could see, to row ashore at the bypass anchorage, pulling the hawsers along. The rest of the day they snoozed, or presided over a little mat full of wares—coins, stamps, cheap figurines, leather, cheap jewelry, postcards, sandals, dates—spread out on the fantail for our consumption. We were under orders to keep them out of the labs or under sharp eyes. Similarly for the 'bumboat': Egyptians who pull up alongside the ship in Port Said to sell shell-inlaid wooden plates, little chests, toy camels, rugs, and such.

A tall Egyptian of distinguished mien introduced himself as a director of the Suez Company Research Lab. His launch lay alongside. Reason for the visit: The *Chain* had kindly consented to operate its ocean measurement PGR though the Canal, yielding a bottom and sub-bottom trace to perhaps two feet. The Egyptian government would return this small favor by granting permission to land on Zebirget, a tiny (1.74 square mile) barren volcanic island in the Red Sea's Foul Bay.

On March 20 at anchor to the canal side just north of Ismailia, the *Chain* was last in a convoy of 46 ships going south, awaiting passage of a 38-ship convoy headed north. From the fantail it was a strange spectacle, a morgana effect, to see masts and superstructures loom and glide silently past behind barren mounds of sand. I looked through a half-open hatch of the wet lab. For endless minutes inky blackness; then, a searchlight appeared and moved on. Next, a building with various portholes, topped by more lights, immediately followed by three red lights in a vertical row. To those familiar with the merchant marine, these red lights indicate the vessel carries inflammables with a flash point below 73F.

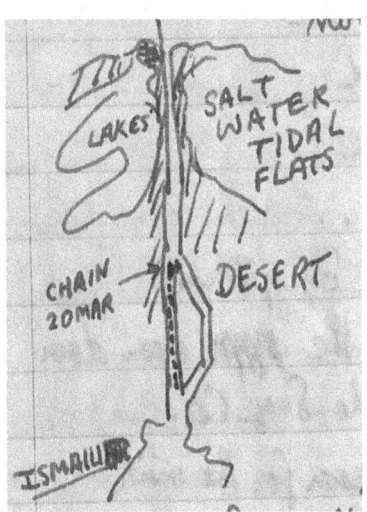

Soon they were gone, but more lights appeared—the illuminated stack, ESSO—and finally the last stern light signaled the end of a supertanker. In the distance a powerful searchlight crept inexorably forward, another floating palace. Ahead, in the south-going fork, a long curving row of light clustered at rest. The next-to-last ship was the crude oil tanker *Kazimah*, from Kuwait. A while ago, their launch came alongside, and a dozen from the *Chain* went back for a visit. Too bad I wasn't around.

At about nine, those of us left behind managed to coax a small, brown Egyptian oarsman to let us row over to the *Kazimah*. She loomed high and empty, but we availed ourselves of a starboard ladder, clambered up and emerged on a vast metal plain, broken here and there by valves and huge pipe elbows. A long, elevated walk ran from the bridge aft to the engine room and quarters. In the chief engineer's lounge (or was it the Conrad Hilton) we found the rest of the gang, hoisting a few drinks and shooting the shit. On the way we filed along a narrow steel walk, suspended over the cavernous engine room, which would be a good setting for a Hitchcock film.

After rowing back slightly high, we were off again for the Suez. I was on watch at sunrise over Great Bitter Lake, part of the canal system. Finally, between the snarling stone lions guarding the entrance, we passed the refinery and convoy smoke into the blue and balmy Gulf of Suez.

PASSAGE 4:
FOUL PLAY IN FOUL BAY

Official Egyptian landing permission had been sought long before to land on Zerbirget, but despite our efforts, it had not been granted. With the Egyptian researcher's trade, geologists on board now eagerly hoped to visit the island.

Also known as St. John's, Zerbirget exposes dense rock generally present only in Earth's lower crust and mantle below the crust. A green crystalline mineral, olivine, also known in the semiprecious gemstone world as peridot and Hawaiian diamond, has been mined there for more than 2,000 years. This ancient history of mining gemstones led to the island's old name of Topazios. Today Zerbirget is part of Elba National Park, a nature preserve with many species of breeding birds. Increased numbers of tourists have impacted the coral reefs around the island.

As we neared, three Egyptian destroyers silhouetted against a rising sun steamed across our bow five miles away. They signaled repeatedly, "What is your purpose here?"—even though our flags indicated RESEARCH. Certainly, their presence here could scarcely be considered a coincidence.

The destroyers, probably comprising the better part of the Egyptian navy, began shelling the island and dropping 50- or 100-pound depth charges overboard, the latter registering as echo-blurring noise on the top-lab recorders, even from such a distance. This was the only answer we got for a harmless, diplomatic-channel request to put a small geological landing party on their four-square mile pile of rock.

> Three months later, Woods Hole Institute got the permission we needed, and on our return from the Indian Ocean some of us visited Zerbirget Island without interference from the Egyptian Navy.
>
> It was June 7 when a landing part of seven pretty well covered Zerbirget. The trip to and from the ship was a bouncy, salty half hour in the Zodiac towed behind the ship's skiff.

Laden down with samples and burnt from the sun, we arrived back in time for the Sunday steak fry. The expedition had aspects of the comic: the overkill gear including a great excess of food (would have outlasted the water), flares, and walkie-talkies for each group. We lost contact with each other after the first half hour.

But carrying my trusty old blue rucksack laden with gear and rock samples all over the island was a pleasantly fatiguing way to spend a Sunday.

[A letter describing what we found was sent to my parents and destroyed with the rest of their house in the 1977 Santa Barbara Sycamore Canyon chaparral fire.]

PASSAGE 5:
CROSSING THE EQUATOR—AN INITIATION

Two thousand miles later, having left behind Saudi Arabia and Yemen for the open ocean, we reached the equator.

April 4 would bring various scurvy, slimy tasks, appropriate to the miserable, scurvy, slimy pollywogs. Big Fritz Hess, a Germanic behemoth, had to do soft-shoe in the galley. Others entertained the honorable Shellbacks, experienced equator crossers, by singing arias from operas. The duties were posted that day, signed HER Majesty.

All this nonsense apparently originated in the US Navy. But Columbus is sometimes said to have started this tradition to keep his sailors entertained as they were often becalmed near the equator. Charles Darwin on HMS *Beagle* recorded an event not too different from modern ones. Some early hazings were rough, causing injuries and sometimes deaths.

April 5, when we actually crossed the equator, was the Royal Court, initiation and barbecue. All pollywogs had to report with unmatched socks and long-sleeved shirts on backwards.

Pollywogs John Cook, a new PhD in geophysics, and I were to report to the boatswain with sponges, water pail, and mop. We sponged down the outside of the winch room, then scrubbed iron oxide stains below the wheelhouse. Thus occupied, we got doused with water from a bucket emptied from the bridge.

At 10:30 p.m., all pollywogs assembled on fantail, sang "April Showers" to the satisfaction of Shellbacks, and got fired upon by fire hose from the boat deck. Then, one by one, we were called to the Royal Court to be summarily tried and duly punished. I was among the first.

> Charges against Vogt: Accused of never having his mouth empty of food, raiding both iceboxes of night rations prior to 1900 every night. Seen walking to the labs with bulging pockets of goodies and sharing them with no one. Beating the chow time by 11 minutes at each meal, eating up all salads and goodies in sight then complaining he is not being fed.

> Conspiring with Friedburg as to who should raid the goodie locker first. Accused of not washing during the first 16 days of voyage. Sentence: Thorough washing and cleaning, inside and out.

Prodded and pushed up, blindfolded, I waited, blindfolded. "Ho, HO! The WORKS!" the judge declared, with raucous laughter.

Then I was smeared good with grease, forced to crawl through a tarpaulin tunnel full of flour and only God knows what. A half dozen photographer bystanders know how I looked.

I crawled whacked along by such vengeful shellbacks as Chief Mate Mysonas, who does the shellback routine with more gusto and relish than anyone else on board. Especially since a pollywog thieved his personally wrought wooden paddle yesterday. And to top things off, Colin Waldon stole his stool, the one Mysonas needed to stand on to read the radar screen, because he is so short.

Next the scurvy, woolly, slimy, miserable pollywog Vogt was given a cleaning, sat down on a horse and sprayed down with the royal fire hose. Now I was ready for the Royal Family: WHOI oceanographer Betty Bunch as King (as I had predicted); Fat Louis as Queen; and Fred, an electrician, as Royal Baby, complete with diapers, cigarette dangling in mouth, bulging biceps adorned with sailors' tattoos.

Pollywog knelt and bowed before Neptunus Rex, Ruler of the Raging Main, and his Queen. Charges were read from a scroll: Stealing goodies from the galley, etc etc. Also I invited a heavy penalty by locking out the guard, laughing loudly while being blindfolded, and otherwise inviting the displeasure of the Royal Court. There was talk of shaving off my beard, what I had feared all along. First, though, I had to kiss the grease-besmeared belly of the Royal Baby, goaded into action by the Queen, touching me with her Royal Scepter, or shall we say her Royal, 36-volt cattle prod. Zap! Kiss the Baby!

On to the Royal Barber. Shampoo: a raw egg bashed on the noggin; a handful of molasses as grease and a cup of flour as powder. Surprise of surprises—they spared my beard, shaving only a three-inch square on the left of my forehead. The Royal Barber did curse my thick, impenetrable hair. The Barbers waxed more liberal with time; later pollywogs lost parts of beards or had heads shaved into landing strips and Maltese crosses.

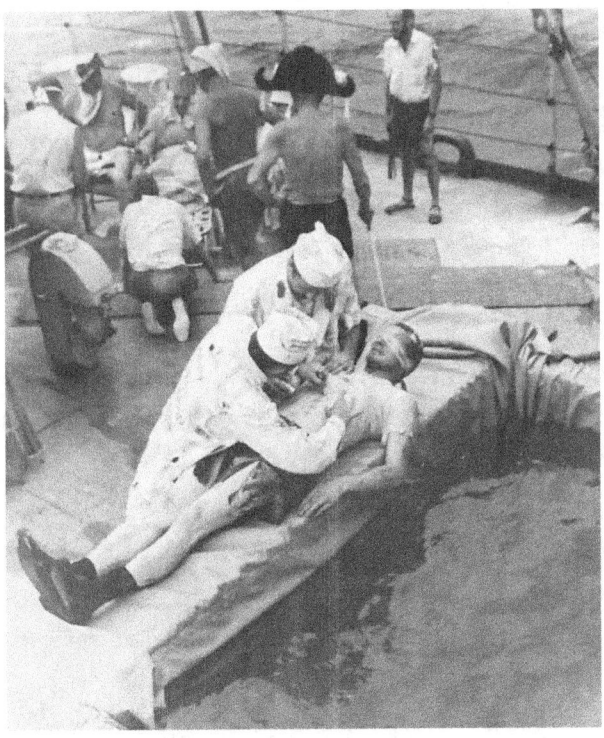

The hapless pollywog. blindfolded and incised with crayon scalpels during the ritual crossing of the equator.

Done with hair trimming, the pollywog passed into the hands of the Royal Surgeon. First a bite to eat: a bad-tasting substance with a clay texture, probably stale bread soaked in vinegar, washed down, literally with vinegar. I gagged and spit out, but I could still taste the vinegar.

The hapless pollywog was then loaded on an operating table, incised with crayon scalpels, and dumped blindfolded into the Royal Bath Tub, filthy from his predecessors. Thus ended the hazing.

I immediately slopped down to the galley and, wet and greasy, embraced the Chief, who was haranguing the remaining pollywogs awaiting their tribulations.

After the last of us had been initiated, the tables were turned, and the pollywogs briefly ruled the fantail. Fritz gunned down the Royal Family with a fire hose. Louie fled, and Fred got pitched into the water. Mysonas was the first to barricade himself behind his doors. One by one the remaining shellbacks were rounded up and dunked, even scientists who claimed such riotous initiations were unheard of on Scripps ships (perhaps because there are so few pollywogs on a given crossing). What a celebrated sight! Only the Old Man remained dry.

Afterwards to taped music on fantail, we partook of barbecue, two beers each, and juicy grilled steaks, framed in a hazy, cloudy, golden sunset.

> I got to experience the Russian version of crossing the Line 11 years later, in 1975, aboard the Soviet SS *Fizik Kurchatov* in the Atlantic. Alcohol was consumed before, during, and after the event. We three foreigners were saved to last—but by then the Royal Bath was filled with the dirtiest water. The captain ordered that Pollywogs (or

what they are called in Russian) not be treated so rough as to break bones, as had happened on other research cruises. A major cultural difference is that King Neptune's henchmen wore blackface and grass skirts, caricaturing imagined African jungle savages. Even in 1964, that would have been unthinkable on any US vessel.

Early in the morning of the next day, our most serious equipment loss occurred. The entire one-half-thus-far-assembled thermistor chain, weighted by a one-ton fish, rattled off a wildly rotating winch. It sounded like a tank going down the street, until with a final jerk, perceptible all over the ship, it disappeared into the briny deep. Toward the end, pieces started flying, and the two observers, Chuck Porter and myself, beat a hasty retreat into the main lab. Ed Tasko, whose new thermistor winch this was, worth $30,000 at least, flailed wildly at the winch controls, shouting into the squawk box for the bridge to stop the ship, then ran for it too. The culprit was a single drive chain link that broke. The other side, being connected by a differential gear, was unable to hold the beast.

PASSAGE 6:
SEYCHELLES—SCIENCE AND LOVELY WOMEN

A long, vibrating rasp indicated we had run aground on coral. A few powerful reverse turns set us free. The fantail was now crowded with *touristas* straining at the leash, for Seychelles was the high point, according to the buildup it had received. Thick tropical jungles clambered up steep granitic slopes and disappeared into a rainy overcast. *Chain* would have no berthing space until the afternoon, so down went the anchor and the first group leaped into a plying boat alongside, for the two-mile trip into Port Victoria.

Contrary to some expectations, no love-starved cluster of ravishing island maidens waited on the stone pier. Only a few curious Black and Creole boys. We caught the dank smell of copra (dried coconut meat), of rotting coconut husks, of muddy flats and low tide, alive with small crabs and a kind of salamander that scurries across a water surface.

No large boats were in the harbor, which looked not much different from the days of the city's namesake. The largest vessel was a 100-foot schooner loading for a trading voyage to Mauritius, that sailed an hour before we did, and in the dusk she could be seen, a full-sailed schooner, to our starboard, with the dark mountains of Mahe behind.

Happily ashore, we trooped downtown. The center of Port Victoria was marked by a smallish ornate clock tower in the middle of an intersection. The clock tower, according to the inscription, was donated in honor of the memory of Queen Victoria by the people and government of the Seychelles. Around the monument were a general post office, Barclays Bank, and some general stores across the street. From there the town stretched along the coastal road (toward the southwestern and distant end of the island) and inland along the road to Beau Vallon.

The roads were asphalt, but branching from them were dirt tracks and trails leading into coconut and cinnamon plantations, and then, as the topography became too steep, into jungle. Practically the entire population lived within 100 yards of the several paved roads, leaving the mountainous interior a wilderness paradise.

A shipmate and I began walking. The more we walked, the more we enjoyed it. A banana-stand man wouldn't accept East African change for his fruits, but he gave us some instead. We exchanged inviting smiles with some dark maidens washing in a stream, their skirts drawn up. We got a scowl from the mother. The road was full of people on foot, on bicycles or packed into open buses—merely trucks with an awning roof. The cars on the island were mainly Morris Minors. A man carried a stringer of fish. Women balanced large cans or bundles on their heads. Little stacks of cinnamon sticks also moved down the road on their human beasts.

Cinnamon is a glossy-leafed bush grown among coconut palms. Every coconut palm I saw on Mahe had a number on it, as high as 500 for the larger groves. Now and then we saw a rickety factory surrounded by huge masses of coconut husks and rotting layers of cinnamon leaves.

We started at noon and about 4 p.m. reached the Tortoise Hotel, where a road crossed the island. The hotel was airy and empty, opened just for us. We drank South African Castle pilsner at 2 rupees (44 cents) for a large bottle. The mulatto girl serving us was interested but skittish. We listened to a heavy, pleasant swish of a tropical downpour in the coconut fronds and glossy breadfruit leaves.

Mahe, the largest of the Seychelles islands, reminded me of Panama: the racial composition, the clothing, the climate and vegetation. Also the brilliant greens of the trees, the white clouds and blue sky, the white laundry washed in local streams and spread out on the grass to dry. The bus which, we were told by an old lady, ran at 7 p.m., never showed up.

At length the skies opened up and we got drenched as never before in a rainstorm. We started to slosh back to Victoria with visibility only a few rods in places where trees overhung the road. Not a car for two hours, just thick, heavy, unrelenting rain. No lightning, and being wet was not unpleasant, but one gets out of shape after lounging around the gravity lab week after week. We got lucky: The first car to come by stopped for us. It was the pilot who had guided the *Chain* into her berth that afternoon.

Peter in flora exploration mode. Mahe Island, Seychelles, 1964

The second day in the Seychelles dawned blue and bright with patchy altocumulus and low clouds over the mountains. The streets were alive with roosters crowing and the commotion of early morning one finds only in the tropics.

After eating, I hiked across the island to Beau Vallon, perhaps five miles, over an 800-foot pass, nourishing myself with bananas and mangos, which I hate. Some boys on the road were slashing up a strange fruit, and I traded a banana for a taste. Juicy, very sweet; they called it *yak*. Perhaps it was a jackfruit.

I came down to the ocean through coconut groves, where untended, narrow-prowed boats and drying fish nets lay on a fair, sandy beach, where I found delightful water for swimming and body surfing, moderate surf compared to California. In the distance, the mountainous Silhouette Island reared out of the blue sea. After frolicking in the warm surf, I wandered down to the Beach Hotel. Its lawn and palm trees ran right down to the average high tide; cyclones were unknown there, as evidenced by the carelessly built grass-and-corrugated-tin roofs.

The Beach Hotel in the community of Beau Vallon was managed by a charming, sexy madame of French extraction and her four teenage daughters. The lithe, barefooted mulatto girls who serve at the Beach Hotel were also something to behold.

Americans were not new here; there were American fellows from a satellite telemetering station that perched improbably on a mountaintop like a huge mushroom.

In addition to ping-pong, swimming, drinking, and listening to the surf, a few of us went skin-diving beyond the breakers at high tide. We found nothing. A heavy tropical downpour commenced suddenly; sea and sand turned to gray, and plants bent under beating water.

I spent the night at Beach Hotel for 20 rupees ($4 full pension).

In the morning, April 15, two shipmates and I hired a shifty little white kid named Patrick as guide. We cut through cinnamon and coconut groves to the stream, then up along the stream, ducking under spiny palms, clambering over mossy granite boulders that were unbelievably slippery.

At first, we maneuvered back and forth carefully to avoid wet feet, but wet feet could not be avoided. The water was so warm it was a pleasure to slosh up the stream bed, pants, shoes and all, hesitating now and then to brush away a large spider (2 to 4 inches across) with a cinnamon stick. They are reputedly harmless creatures but are found in great numbers spinning webs in trees and between telephone and electric cables.

At noon, bathed in sweat and not too clean, I left behind my watch and camera and plunged into a small rapids pool, clothes and all—what a luxurious experience—just below a small aqueduct. I walked back to Victoria, hardly conscious of wet clothes, stopped to buy tortoise shell stuff, had a beer and boarded the ship.

PASSAGE 7:
ROUGH SEAS, FEVER DREAMS, A LOST DREDGE

On April 19, crew member Stires got a steel splinter in his eye, which affected subsequent research plans. As he needed medical care, we changed course for an unplanned first stop in Mauritius, dropping Stires off for hospitalization.

We steamed into and out of Mauritius while I was still asleep. By afternoon the low mountains of that island, shaded by billowing cumulus, were far astern.

A man-o'-war flew overhead, and away to the port a small unidentified freighter steamed rapidly by, the first ship in almost a month in the open ocean. The Indian Ocean is a shipping desert. We steamed north and easterly to the abyssal plain, from where we hastily turned for Mauritius.

On April 28, we attempted a pipe core in rough seas, over 200-fathom (S.M. Ridge). Due to ship drift, the wire got fouled in the screws and had to be amputated with an acetylene torch. Luck has been against us since Stires got the steel splinter in his eye.

In the early morning of the 29th, on my watch, we caught part of an Indian Ocean cyclone, 40-knot winds and heavy seas from the south. Seawater gurgled and sloshed, competing with air for passage through the scuppers. The ship lurched and shuddered as it collided with huge water masses. We were barely holding our own. At 6 a.m., the bridge was forced to point the ship into the seas.

From the deck of RV *Chain*, 40-knot winds and heavy seas.
North Atlantic, 1964.

Shortly after noon, I was almost spilled out of my bunk by violent rolls; a heavy unidentified object slid off the air conditioning duct, clobbering me on the leg. There was great clattering and screeching of loose objects and things straining to be free from the call of gravity. According to Ziggy, our German guest geologist Siegfried, ketchup bottles and the captain suddenly found themselves along the port bulkhead of the galley. The imperturbable master had a concerned look then, and a few minutes later the bridge changed course to 330. Instead of fighting it out, he decided to get out of the way. We steamed 330 with following seas until the next morning, then shifted to 180 bound for Mauritius.

On May Day, we managed to take a heat probe despite the rather turbulent seas. The 10-to-20-knot winds we were getting were the combined effect of a South African high and the storm moving southeast at 10 knots. That roll, by the way, was 35 degrees, the record for this trip.

I sat it out in the gravity lab smoking a pipe and prying chunks of coconut off the shell as Aldrich sang, "Ah don't care if it rains or freezes, s'long as ah got my plastic Jesus," while Porter and Aldrich told stories of stealing tombstones.

Caught by some virus, I slept 15 of 24 hours. Fevered dreaming extended back to high school days in Ohio: insecurity, fear of scholastic and professional defeat, defeat of the ego. Sometimes I give up all hope of writing on expedition data, whereupon the question arises: Now whither? How do advanced degrees figure in my scheme of life? In truth, I have made no overwhelming effort to impress the higher-ups of my scientific motivation and competence. The next two weeks will decide, finally, how this rather luckless expedition will pan out for me.

On our second passage to Mauritius we landed and I found an overwhelmingly friendly island. The day before we sailed I took a grand bus tour around the southeastern end of Mauritius to Point Louis, Bambou, La Marne,

Souillac-Mahebourg and Curepipe to Port Louis. It took about six different bus lines and changes to accomplish this. The conductors were unbelievably friendly. I asked and answered questions with great curiosity.

I climbed on the steep volcanic bluff near Beau de Cap, where huge breakers smash on reef corals only 1,000 feet from shore, forming a fairly deep, calm lagoon for coastal traffic. At Souillac, the surf approaches within several hundred feet and even breaks on the hard stump of an ancient volcanic pipe, sending up great clouds of white spray. I swam in the calm water, 75 degrees or so, at Telfair Garden, a small roadside park where tongues of coral-sand beach lap around the worn stumps.

On the road from Mahebourg to Curepipe, a stranger might have been baffled by a number of large signs reading DANGER: BLACK SPOT, a warning accentuated by a solid black circular spot. Road construction, perhaps? Or a diseased area?

No, the Black Spots were areas where a white man could not walk alone without fearing for his life, especially at night. An Englishman was found one morning, barely alive at the roadside. His arms and legs had been amputated and his testicles stuffed in his mouth. I've heard that the local populace had been aroused to wrath because of the careless way white drivers screamed around the curved roads. Frequent mists made driving difficult and dangerous, and a number of locals had apparently been run down; hence retribution. There also may have been a branch of a notorious Indian death-worshipping cult at work, which certainly made a better story. (This Black Spot business might have been a good setting for an Alfred Hitchcock plot.)

Back at sea, on May 13, another dredge was lost. We lost many rock dredges, but dredging rocks from the seafloor was not unlike fishing. If the rocks were not loose or easily broken off, the dredge either came up empty or was snagged and trying to pull it loose often caused the necessary weak link just above the dredge to break, so the cable was spooled back onto the winch and the dredge was donated to Davy Jones. We were trying to get samples of igneous rocks but lost most, if not all, the Indian Ocean dredges on reef-type carbonates (basically limestone) cemented together. However, *Chain* was well prepared with machine tools and stock metal for building new dredges, and other stuff, onboard.

Over the squawk box a lively Scottish dance scratched and faded, improving my morale. Surprisingly it was often the skipper who beamed these little musical gems into the intercom. At 5 in the morning a bit of beat from far away does the heart good. The 4-to-8 watch is also known for the crossfire of witticisms between the bridge, and the top and main labs. During those hours, most of the ship sleeps..

The other night scientists, Aldrich, Carl Bowin, Fritz Hess, and I sat around shooting the shit. Topics ranged from Rosicrucianism to the exact definition of Immaculate Conception to Disneyland's Matterhorn, and ending in a raucous monologue by Hess on the subject of the chief mate, Mysonas. Hess hates "the little bastard" and has infuriated the shell-shocked WWII veteran by setting off firecrackers at unwanted moments.

The other day Powers and I stood just inside the port fantail hatch. He dipped a firecracker into my pipe bowl, opened the door, threw it out, and we returned to the computer in a jiffy. It turned out the chief was standing at the bottom of the steps that come down from the boat deck when it went off. Fritz Hess was immediately suspected, and Mysonas tore off after the supposed culprit, in a rage. Hess enjoyed the situation and said nothing. Last night we plastered the main lab bulkheads with ingenious Hate Hess posters, with a shit list that included everyone except the gravity watch. Today the captain sauntered by, stopped, and without a word ripped off all our great wit.

PASSAGE 8:
THE EQUATOR, AGAIN—AN EXISTENTIAL
MOMENT AND A WHALE OF A SHOW

On May 19, at 6 a.m., a great golden sphere rose out of the eastern sea. Scorpio and Sagittarius were high in the western sky, and the bright galactic center was surrounded by dark dust. Last night's sunset was perhaps the most spectacular I have beheld; the atmospheric visibility was perhaps 100 miles. Large fleecy cumuli with half-mile ceilings appeared to float on the sea. After the sun had sunk, the western edge was spectacular enough for any mortal; the still blue sky became streaked beyond the zenith with great golden rays, clouds far beneath the horizon throwing their shadows onto the stratospheric dust. The great, flaring crown lasted long after sunset.

The sunset followed a cookout on the bow as the heat probe was being winched out of the abyssal depths off the Seychelles.

After the evening watch, I watched the moonlight, a dappled highway down which the *Chain* steamed. Arias from *Madame Butterfly* and jazz, "tantalizing music" in the words of Al, the only African American aboard. This middle-aged engine room man knew more about the sea than many of our instruments. From the color of the water he knew when we were up on a bank. Al spent most of his spare time fishing. Yesterday, on station, he and Powers hooked a huge, gray 8.5-foot shark, which was hoisted aboard with the hydrocrane for the purposes of decapitation and jaw extraction.

On the morning of May 29 we crossed the equator again, at 62° E. Having concluded the magnetic anomaly strike studies, the *Chain* was steered to the northwest in the direction of Socotra. During the afternoon we had a small initiation into Neptune's Court for new pollywogs.

Bowin had to serve lunch to the crew, while Ungar served us. The costumes were the same as before. The king was naturally singer-rhetorician Aldrich, and the fat prankster Feden mercilessly prodded the helpless with 36 volts as Neptune's Queen. The colossal Germanic Fritz Hess was the Royal Baby, whose enormous fat-smeared belly had to be kissed three times by each pollywog. I came as an uninvited Royal Mexican wearing a tent hat that I purchased for 5 rupees in Victoria and a colorful blanket I had purchased in Morocco.

On June 2, the hazy mountains of the African coast were visible most of the afternoon. No trace of the monsoon inside the island of Socotra, off the Somali Peninsula; only a light northeast breeze rippled the flat sea. Blue sky merged with continental summer haze, and an isolated thunderhead reared out of the dusty air far away over Africa.

That night, the Southern Cross twinkled low over Somalia and the warm air whispered of seacoast and savannahs. The bow wave twinkled merrily in a constant cascade of phosphorescent organisms. An unknown ship overtook us on the starboard beam. Liberty Valance was getting shot onscreen in the galley.

The night of June 3 found me on the bow, sitting in a folding chair sipping coke. BAF British Armed Forces Aden, loud and clear on broadcast, brought light music and news from London, including the nauseating news that Barry Goldwater won the California primary. Faint flares of heat lightning, almost as far removed as American politics, illuminated the hazy distance over Somalia. The mountainous coast of Africa had been in view almost continuously since Cape Guardafui, the Horn of Africa. An occasional meteor whisked into oblivion, and no sooner had I espied what was probably an Echo satellite drifting serenely over the Great Bear than it passed out of sight into the Earth's shadow.

Try as I may to forget, my daily sleeping pattern seemed continuously to drift out of phase with the sun. I had the 4-to-8 a.m. watch for over a month, mainly out of deference to colleagues who seemed to be creatures of greater habit. I had been sleeping less and more poorly each night, then spending the day catching up, to wake up an hour or less before the afternoon watch. Before watch, I usually spent some time on the thermistor winch, inviting skin cancer and sunstroke.

On the subject of cancer, I had been subject to slight, rather vague pains in the right chest. Lungs? Last night, I leaned over the bow staring toward the murky horizon of the Gulf of Aden. The fatalistic yet statistically disquieting possibility occurred to me that I would die of cancer. As if to dissuade such thoughts, a rare wave sent salt spray into my face. The last few days had seen me stalking the ship like a hang dog, sapped of enthusiasm and vitality.

Yesterday morning, the morning watches were treated to a whale of show.

"Thar she blows, the Great White Whale," announced the squawk box, and all rushed out on the fantail to see what was up. Sure enough, 1,000 yards abeam to the port, a periodic geyser erupted from the flat sea into the morning haze. Another whale, and another. An hour later, whales were seen again, one fellow within 500 or 1,000 feet of the ship, humping majestically for several seconds, accompanied by a snort of steam, to submerge in a roil of foam. A dark brown color, perhaps 30 feet or more in length. Still more whales spouted off the port just before breakfast.

Five in the afternoon of June 4 found several dozen chaps on the bow, leaning over the rail, their butts bobbing with each resounding cheer. The cause for this commotion was a platoon of porpoises riding the bow wave, in formation. Three in the upper echelon, three in the lower. The two large porpoises were only inches below the glassy blue water, and every minute or so apparently out of sheer exhibitionism, one of them would hurtle out, only to plump back with resounding belly smackers. Whereupon the audience would cheer. The last performer would often roll on its side to take a look. A small clever porpoise would often make quick, darting rolls over the back of one its large neighbors, also drawing cheers.

At such times one is moved by uncanny feelings of warm mammalian comradeship towards these creatures. For every instance when they are actually off the bow there are many when their cheerful chirping is overheard by the echo sounder.

PASSAGE 9:
MALLARDS PADDLING PAST STONE LIONS

Some miles off Suez, we steamed into the smoke-filled outer staging area. After an hour at anchor we changed over to the inner staging area to join the waiting convoy. The *Chain* anchored at the 50-yard line, so to speak, with ships of the south-going convoy moving by in succession.

There was the usual majority of Persian Gulf tankers. On and on they came, through the afternoon and evening. A whole fleet of Russian fishing trawlers emerged, gray tugs with red stacks and yellow hammer and sickle, like a file of mallards paddling out past the stone lions, one, two, three...15! The earlier ducks waited up until the flock had reassembled, then off they went.

Our fantail became a colorful bazaar.

Our fantail had become a colorful bazaar. Inlaid plates, toy camels, camel rugs, hassocks, sheik head-dresses; much bargaining and doubtless many bargains. A sarcastic, one-legged Arab mumbled about "fuck-books" and pictures. On the sly (don't let the Captain see!) he produced poorly printed paperbacks from France, with unseparated pages.

The sun was hot, the inversion evaporated, and a cooling sea breeze swept the convoy smoke away. Arid sedimentary escarpments stood starkly in a light blue sky. At 6 p.m. the onshore breeze suddenly yielded to a warm, dry desert air; the thermometer climbed to 90 degrees or so. The next morning was almost totally calm; again plumes of smoke spread their brown wings into the stagnant atmosphere (no smog, though!).

After all the tankers pulled anchor and steamed in, the freighters went, led by the *Schwarzenfels* from Bremen. We were near the end of the convoy. In Great Bitter Lake, we dropped anchor to wait for a number of south-going ships to come out of the next one-way stretch. Great Bitter Lake had two-way traffic with the ships only a stone's throw apart.

The lake had the color and probably the temperature of good pea soup. It might well have been a summer day in Oklahoma; the temperature rose to 97 with hot dry winds and some high, light cumulus. MIG fighters stood in the sun near the lake. Top Lab called down one-kilometer canal markers, and a slide rule computed speed.

We steamed serenely into the commotion of Port Said, like a huge house trailer moving down a 15-lane super-highway, pre-empting the whole road. Great tankers were tooting and sputtering and honking, and small craft scurrying. Now and then the stately snort of a great ship sounded above the din. Our cherry picker dropped the little Arab skiff, our pilot and the 'high-ranking visitors' from the Canal Research Institute into launches pulled alongside. The *Chain* relinquished her canal beacon and bid adieu. Cool breezes and a light chop welcomed us back into a more familiar ocean: the Mediterranean.

PASSAGE 10:

IN LEBANON, A BUS FOR BAALBEK

I visited the famous Roman Baalbek ruins during our Beirut port stop.

Coming into Beirut, a sprawling white city backed by high mountains with patches of snow still glimmering in a midsummer sun to the northeast, we tied up behind the 30,000-ton Russian factory ship *Slava* on the early morning of June 12.

After hours of impatient waiting for the customs and health clearance, we prowled around the city around midday. It was a loud city, especially after the tranquility of the islands.

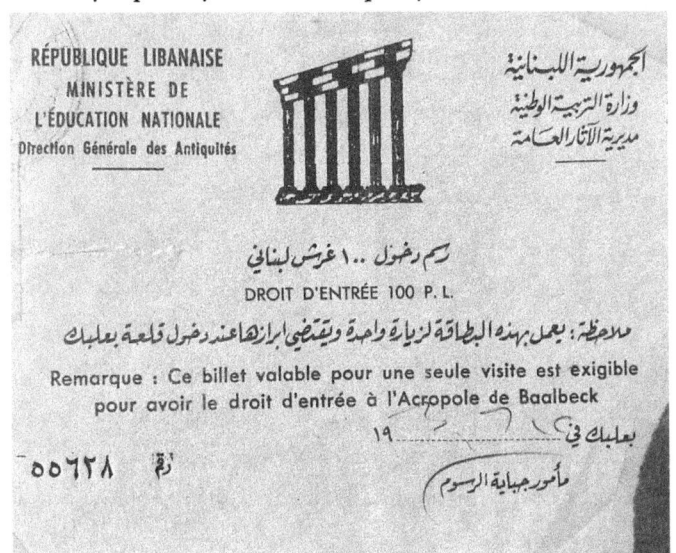

RÉPUBLIQUE LIBANAISE
MINISTÈRE DE
L'ÉDUCATION NATIONALE
Direction Générale des Antiquités

الجمهورية اللبنانية
وزارة التربية الوطنية
مديرية الآثار العامة

رسم دخول ..١ غرش لبناني
DROIT D'ENTRÉE 100 P. L.

ملاحظة: يعمل بهذه البطاقة لزيارة واحدة وتقتضي ابرازها عند دخول قلعة بعلبك
Remarque : Ce billet valable pour une seule visite est exigible pour avoir le droit d'entrée à l'Acropole de Baalbeck

بعلبك في ١٩
مأمور جباية الرسوم

رقم ٥٥٦٢٨

Carrot juice, freshly made, was sold at stands, along with peppermint and onion salad, shrimp wrapped in apricot leaves, and other goodies. Prices were almost American. The proportion of women to men on the streets was more than 1- in-10, compared to 1-in-100 or less in a place like the Arab section of Aden or a provincial town like Baalbek. Some striking women; many had rather fat legs. The pale skin colors of many Arab women, of dark complexion in a sunny climate, meant just one thing: They rarely got outside.

In Beirut, the Mercedes had almost totally replaced the mule. Such prosperity must have come from trading and banking. Agriculturally, there was not much: mountains, stony and thin soil. There were vineyards and grain in the hotter, drier valley to the east. The coastal valleys and hills, too steep for terracing, exhibited a California-type brush vegetation on mainly carbonaceous rocks. The mountains above about 4,000 feet were barren grasslands, presumably covered by conifers in earlier days. A few celebrated groves of cedars remained. A ski area was located in one of them. A good idea for a vacation would be to ski those out-of-the-way places: Norway, Yugoslavia, the Atlas Mountains, Lebanon, and Iran.

The Lebanese high country was dry, grassy, and pleasant. I boarded a local bus for Baalbek the same afternoon. Bus proprietors hawked their destinations quite vociferously so there was little danger of boarding the wrong bus. Fare to Baalbek: 2 pounds, or 66 cents. The bus ground up through endless suburban apartments. My ride was aggravated by a pesty kid who continually spat fruit seeds out the window in front of mine.

Beyond the Lebanon Range was a flat valley much like that around Santa Ynez, California: clear, warm (perhaps 85-90 F), and dry. A few light cumulus drifted along. Not a poor place for man to begin with the long task of civilization! Water came from mountains and limestone springs. The city of Beirut was supplied from a Stygian reservoir coiled around massive dripstone formations.

It took three hours to get to Baalbek. No pre-Roman remains survived. But outside of Rome itself, I doubt if any classical ruins outshone those at Baalbek, from the point of sheer massiveness and aesthetic appeal. Over the vast complex of temple ruins towered the last six columns of a mighty edifice dedicated to Jupiter. These columns, like sequoia trunks, were visible from far away as one approached the city. The temple was mainly limestone, although a number of warm pink syenite columns gleamed in the dry middle-eastern sun. Hollyhocks bloomed among the crumbling stones.

Near the outskirts was the limestone quarry used in classical days. In it, half submerged in succeeding sediments, perhaps the largest block of rock cut out in pre-modern days, perhaps 100 feet or 200 feet from its origin. Moving such a monolith would have required 40,000 slaves.

No army of 20th century tourists pounded on Baalbek's gates on this day. (In July and August, visitors would flood the ruins briefly, for some great staging of music and drama.) Here, one could still explore the crags and crannies in peace; no ubiquitous guards and signs. A small boy unblocked a hidden door; a pitch black spiral staircase of stone led to the precipice atop what remained of the Temple of Bacchus. One could edge along the tops of columns, peering respectfully through yawning bathtub-sized voids where the vaulted roof had collapsed.

Framed by Roman columns and straight green poplars in poetic juxtaposition, the smooth heights of the Lebanon Range rose through the clear air, still brushed carelessly with a wintry white.

An Arab on an ass plodded through the dust, his transistor radio jarring the peace with our century's sounds. Arab tots still pranced behind the stranger, bestowing him with a nettle to test his stupidity, or a poppy for his lapel.

On the second day, I had planned to visit Damascus. But visas were only granted to those "of Christian Faith." Not being able to locate my birth certificate, I chose to spend the afternoon in Byblos up the coast, a half hour or so by bus. This fishing village, which had always gone by a different Arab name I can't remember, was excavated entirely by the Lebanese Antiquities Department.

On a small hill near the water, Phoenician temple foundations overlay Neolithic graves, topped by Roman columns. A powerful Crusader fortress, nearly intact, crowned the historic plot. In its walls, stones cut by earlier men were used again. Of the Phoenician ruins, a small temple foundation with 10-foot-high obelisks and a ship's anchor remained, far more than had been scraped up at Tyre or Sidon. Most impressive were a number of Bronze and early Iron Age graves for local kings. A 20-by-20-foot shaft was sunk into solid rock 40 feet down, and a massive, simply styled sarcophagus somehow lowered into the depths, perhaps by filling the shaft with sand and then mining out the sand, as has been suggested. The archaeologists excavated a tunnel now open to visitors, leading to the sarcophagus in which the remains of Bronze Age King Abishemu of Byblos had slumbered undisturbed for nearly 40 centuries. The sarcophagus, bathed in dim, eerie light and surrounded by moss and small ferns, strikes one suddenly with a sense of time.

Dark but rainless low clouds gathered against the mountains as the late afternoon sun sparkled on a choppy sea. A stiff breeze blew on shore, and short-term Mediterranean breakers framed among the ruins of a Medieval harbor fort by the Well of the Princess, where local people still panned the sea-sifted sands of Byblos for coins and trinkets of antiquity. An antique shipman was willing to sell me 110 non-precious coins (mainly Roman) for $60. As I had neither the understanding of such coins nor the money, I didn't buy them although there may have been the opportunity for some profit.

The evening I spent in the warm company of Lebanese brandy, staggering among the quays. There was a Russian passenger liner, quite a trim ship too, the *Armenia*. When it sailed, only a handful of elderly Armenians waved goodbye to their free relatives here in Lebanon. What a tragic sight for a fancy ship with colored deck bulbs, music playing, deck chairs, and a swimming pool.

A Lebanese guard in front of the next ship, the Greek passenger liner *Messalia*, mistook me for a passenger. I got on board and wandered around. Asking for the bar, I ran into a dark-eyed, smiling Greek beauty with whom I couldn't communicate. I did discover the bar was closed before they threw me off the ship. The next day brought a hefty hangover.

PASSAGE 11:

RETURN TO ITALY WITH STROMBOLI,
CALYPSO, ANTICS ASHORE

More than three months after her first visit, *Chain* was once more on her way to La Spezia. One of the ship's generators had been acting up, and we expected a stop for parts.

From the Mediterranean, we passed through the Straits of Messina, headed toward the island of Stromboli, with the top of the top lab like Santa Monica Beach on August afternoons: binocs, radios, paperbacks, hammocks, folding chairs, cameras.

There being a few hours left, we put in a last station small core and heat flow. The core barrel slammed into hard rock, but the show was worth it. Practically in the shadow of Stromboli, barely outside the 6-mile limit, we could see a slight smoke plume in the north, as the steep, ash-strewn cone emerged out of the sea.

Passage on ships affords time to advance one's artistic horizons.
Chain passes through the Straits of Messina.

At the station just to the south, night had descended and it became clear why Stromboli is known as the Lighthouse of the Mediterranean. Vulcan's fires burned in two places near the summit, like distant forest fires. Stromboli seemed to wink and glare like the eyes of a monstrous toad. The left one was usually brighter. It tossed out a (100-foot-high?) geyser of glowing bombs and cinders at irregular, perhaps 15 minute intervals. With glasses one could follow larger glowing fragments bounding down the upper slopes before cooling. On the other side, larger volcanic islands—spattered with snowy ash and surrounded with bizarre, sea-cliff erosional remnants—had disappeared into the night. Among them was Vulcano, his anger so pent up that his fires sleep for 20 years between explosions. We sat on our fantail eating, listening to radio, enjoying the flat sea and mild night as the trawl winch whined away.

A ship lit up, and apparently drifting, attracts attention. On repeated occasions nearby vessels have altered course for a look-see. When they see the red-white-red of gear overboard or the name *Chain*, a quick glance at a ship register and they steam briskly on. On the night of June 26, a huge tanker passed within one-half mile, forcing a temporary halt to core lowering, which was about to get underway. Night prior to last I was engrossed in a broadcast on the after deck when a small freighter appeared from nowhere, charging past at 200 yards, a black silhouette in the Moon's silver trail. In a few minutes it was only a point of light on a black horizon. Often ships mistake us for a Navy ship, which accounts for a lack of friendliness (e.g. semaphore). Buzzing by military planes is no rare event, and several days ago a US destroyer also came around to have a look.

We returned to La Spezia at June's end. Tied up alongside *Chain* was Cousteau's *Calypso* (Toulouse), much smaller with a crew of 25 and a yellow diving bell on deck.

After months at sea and in distant ports, La Spezia was a relaxed stop: first day, a Portovenere swim; second day at Viareggio reminiscing and swimming; and the third day alone at Portovenere again. I rented a boat and rowed along steep cliffs and cave-like overhangs around Byron's favorite grotto below the ruined castle and the Romanesque chapel of Portovenere. The slurp and slosh of the Mediterranean among the crevasses, girls sunning on rocks, and a bright warm sun that sets after eight, a ripe orange ingested by hazy distant cliffs.

On my first visit, the spot was crowded with holiday Italians, deserting the cities on St. Pietro's day, for the chapel was in fact the Chapel of St Pietro. Below it a peninsular fist of semi-metamorphosed limestones thrust into the blue depths of the sea. The tourist girls from Anglo-Germanic lands I had expected to meet around here and Viareggio must have gone elsewhere.

But: "So much I wined I didn't mind."

July 2, *Chain* was doing coastal work south of La Spezia, coming within sight of Corsica. When we go back depends on the arrival of parts. Scuttlebutt was sweeping the ship. Apparently the main drive generators, which should have infinite ground resistance (armature windings to shaft) were down to 10 megohms in Beirut, presently at 30 kilohms. Some said we had only a 25 percent chance of making it back to Woods Hole without extensive repairs.

July 6 dawned hazy and warm, with massive dark clouds in a wall northeast through southeast and occasional lightning. A noisy magnetic record coincided with electrical activity. I was in the wheelhouse trying my hand at steering the ship a while for kicks. On the radar, 40-mile radius scale, the northern part of Corsica and the peninsular extensions near La Spezia were clearly visible near the edge of the screen. Running along the latter coastline was a fine example of a linear squall line perhaps 20 or 30 miles long; indeed most of the lightning came from that direction. All was what is commonly referred to as heat lightning, although nothing more than distant discharges obscured by interesting cloud banks. In contra-distinction to the sparker noise, this electrical storm noise almost always diminished the total field, by 30 to 300 or so gammas. We were on a course to pass within five miles of Jacques Cousteau's Bouee-Laboratoire, a floating island anchored at great depth, manned year round.

On the morning of July 10, the seas were 10-to-15 feet with strong westerly winds. The weather was sunny and clear with deep blue water and foaming whitecaps. We got into La Spezia once again about noon, and there was shore leave until four, when they decided to spend the night until the seas improved.

Aldrich and I took a bus to Lerici, climbing around the castello, which made a romantically situated youth hostel. We ate spaghetti and drank wine in the jovial company of English, Romans, and Viennese.

Later we found two English girls with whom we had pizza and more wine. Then we met still more hostelers, a shaggy Aussie, a North Island Kiwi who had attended the University of Canterbury, and three charming girls from the University of Toronto (Trinity College), one of whom I fell for immediately. I barely had time to take her back to the castle door, which was closed. Goddamnit though, a guard did open it for her. Curse the curfew. It all went so fast I felt like smashing things. Aldrich, who never touches so much as a drop, prevented me from falling off the castle at various places.

Our hulking vessel stands out among millionaire yachts. Monaco, 1964.

Back on the ship, I kicked a card table and cards across the day room, made a lot of ruckus in our cabin, and hardly slept all night. I threw up some of the afternoon's eatings, which included coffee, ice cream, sweet roll, eight raw mussels, spaghetti, pizza, and large quantities of wine. Those blue mussels are eaten raw with lemon juice, cost 20 lire apiece and seem to be the ones used for bait in California.

At dawn, I got up and walked around the harbor. It was a cool sparkling morning and the hills were clear, so much like the Santa Monica Mountains in April after a rain. In two days, we would be in Monaco.

July 14 was Bastille Day and I was off for Nice, a 25-minute trip, where I watched a military parade with soldiers in berets marching. The parade ended with fire trucks and a truck full of sea-rescue scuba divers. Thousands of holiday French and vacation tourists milled along the long promenade-avenue and the pebbly beach of Nice.

PASSAGE 12:
ALONG THE STRAITS OF GIBRALTAR TO THE CORNISH COAST, A VOW OF TEMPERANCE, AND CREEPY MODS

On July 23, as we steamed southwest of Spain before dawn, we were buzzed by a low-flying search-lighting Air Force radar seaplane. Hours prior to Gibraltar, the water had turned a familiar green; surface and temperature at 85 foot depth plunged erratically upward and downward as we got into inflowing Atlantic surface water.

As we passed through the Straits of Gibraltar, two-mile visibility precluded sight of the Rock. In hazy sunshine (very hazy, including California-type ozone), porpoises were playing all around, some within 25 feet of the fantail. Blackfish, sunfish, and various birds were abundant. We made many close passes as huge freighters or tankers emerged from the haze a mile or two away. We exchanged friendly horn blasts with a Russian scientific (missile tracking?) ship. Thermistor chain tests continued; towing the whole chain. Later, the mountains of Spain were faintly visible through the haze to the north.

On July 24, we steamed northwest off the coast of Portugal, clocks still one hour ahead of GMT summer time. Water pitched into long swells from the northwest since yesterday morning. Under low stratus clouds with drizzle evening and night, the fog horn blared away every minute on the minute. The thermistor chain leaked and salt water entered some of the many connection plugs, in spite of apparently tight fit and liberal silicone grease and taping. Also, the sparker seemed to trigger the thermistor units, which is hardly desirable.

We passed within two miles of Eddystone Light on July 28, entering the Bay of Plymouth under chill westerly winds, clouds at various levels. On land, we could see green fields and rolling woods, castles and smokestacks.

> *My father was the keeper of the Eddystone light,*
> *Slept with a Mermaid one wild night.*
> *YOh ho ho where the winds blow free,*
> *oh for a life on the rolling sea.*

Girl-wise, England was a disappointment. The best-looking English girl was a sexy blond in short tennis shorts, her long legs drawn up against her as she relaxed with her tennis racket and a glass of milk. She sat on a little island of sky-blue billboard, which bordered sharply on the gray-brown brick and the gray English sky. This bit of sunshiny sex represented not the British Government Tourist Office but a concerted (and successful) effort of the English Dairy Industry to woo John Bull from his ale flagon. "DRINKA PINTA" the caption read, coyly, and seductively.

English ale is weaker than water and impossible to get high on. After the pubs closed at 10 or 11, I found lodging at Saumpy Place in Salisbury, after all but resigning myself to a clammy night on a park bench. The proprietress was a gregarious Irish woman, and the pub reverberated with fine amateur Irish voices. I arose afflicted with a suddenly appearing, persistent pain between my lower gut and spine. Kidney? Liver? I resolved to lay off this immoderate drinking I had done during these port stops.

The next day, I rented bikes with shipmates and straggled and struggled around mellow, summery English countryside. We stopped for a ploughman's lunch and ale in a village outside Salisbury, visited Old Sarum ruins and heard organ music in the great Salisbury cathedral. On the streets there were a lot of creepy mods and rockers, foolish but apparently they were harmless.

On August 1, we steamed south, 100 miles from the Cornish coast, on a shelf survey, continuing WSW over the Western Europe Basin toward the first velocimeter station at 14° W.

Two days steaming would take us to the Mid-Atlantic Ridge, where we planned a series of six close-spaced (5 or 10 miles) heat probes in the Median Rift valley and dredge stations on the flanks of the ridge. We had 15-knot westerly winds, 8-10 feet seas and light rain, the first, except for drizzle, in a long, long while. The last rain at sea was a week north of the Seychelles.

In the narrow Median Rift valley, the last probe attempted collected 10 feet of gray mud core, but the probe hit a rock at that depth, thoroughly jarring the instrument.

The dredge haul from the lower slopes adjacent to the valley was spectacular: a bushel basket-sized load of cobble-sized basalt, a chunk of hard limestone and some brown tuff with small, dark, glass fragments. Then a log floated by, and it was duly hauled aboard with a cherry picker, all bristling with gooseneck barnacles, a little lode of sea life-crabs, jellyfish. Several 10-inch sea bass lurked under the log. Colin managed to get one with his spear gun.

Later, during that afternoon's survey, several dozen small whales—blackfish—appeared, snorting and humping around the ship, sometimes as close as 150 feet. Even the oldest salt on board came out for a look, and cameras clicked. The whales milled around in a tight school for at least 20 minutes before slowly arcing off.

The valley here lay around 1,400 fathoms, and ridges rose steeply to above 500 fathoms, in places towering a mile above the narrow, 1-to-2-mile graben, an elongated block of Earth's crust. The floor itself was not flat but broken by a low central ridge in places. A 400–800 gamma+ anomaly mirrored topography, being nearly zero over the valleys. Bouguer, a gravity anomaly corrected for the height at which it was measured and the attraction of terrain, had almost as much relief as free air, dropping only slightly over the rift valley.

During the last week I was engrossed in the main lab, playing back echo-sounding tapes from Seine Bank, Messina, Gulf of Suez, Somali Abyssal Plain.

PASSAGE 13:
IN SIGHT OF HOME, SILLINESS SETS IN

On Aug. 13, the long-awaited course change—*Heading Home*—was announced. No more stations, no more surveying. In short order, the mood on board rose to exuberance: 150 hours to home. Constant outbursts of wit and obscenity, mock vomiting and assorted barnyard sounds over the squawk box. In the evening there were European stations: French, English, Spanish, Portuguese. In the morning, the stronger radio voices from the New York area reached out to us on AM bands. So relaxed and pleasant.

There was a flurry of trading funny money and curios. I traded four of my five Roman coins to Bob Feden for his smelly, badly treated, gross zoological curiosity: a Seychelles land tortoise. Bob then traded two of the coins to Friz Campbell for an Egyptian shell and wood inlaid box (bought for $1).

On August 17, we steamed 270 at 13.2 knots over the monotonous expanses of the central Sohm Abyssal Plain. Steerage was lost for several hours this evening, and I tried my hand at manual steering—immediately getting 30 degrees off course. It requires leverage and body weight, practically standing on the wheel spokes to get it around.

August 20 marked six months out. Total steaming: 30,000 miles at the end of cruise. Total consumption of eggs, due to run out today: 30,000!

A day later, we hoisted a homemade banner, *Homeward Bound from Indian Ocean*, strung between the radar mast and the main mast. At 8 a.m., the cacophony of foghorns, sirens, a bagpipe, and a tooting band sounded under the banner, suspended with a pair of pinkish red pajamas, amid cheers. On a warm morning, as sunlight broke from between clouds, a crowd of some 200 people gathered at the Woods Hole Institution dock. We had to endure an anticlimactic wait before customs clearance.

I walked off *Chain* loaded with souvenirs, including two I'm ashamed of today: a badly preserved Aldabra land tortoise from a trade; and a polished and well-taxidermied Hawksbill sea turtle. Both species are globally protected and I would have been—rightly—arrested bringing them back to the US today. The Aldabra tortoise deteriorated in our home through the years so I paid a Maryland taxidermy expert to repair it. It was his first and likely only land tortoise. I donated both specimens to the Calvert Marine Museum. Paleontologists often refer to complete examples of modern relatives when they work with fragmented fossils of related and perhaps ancestral types.

When I returned to Madison, Wisconsin, I discovered that the girl I had been dating was now going with another guy. Imagine that: She could not wait seven months. But her loss was my gain: I would meet the girl who became my wife a year later.

CHAPTER 3

Blue Floes, Gray Skies, White Bears and Red Navy: US Ice breaker Geoscience and Geopolitics North of Siberia

In 1965 and 1966, I joined two icebreaker cruises exploring the same region. Though I had minimal input to survey design, the 1965 *Northwind* and 1966 *Atka* cruises provided me with the lion's share of geophysical data for my PhD thesis, a reconnaissance investigation of what lies below the Kara, Barents, Greenland, Norwegian, North, and Labrador seas, and the Arctic Ocean. I traveled probably 50,000 miles, as if twice around the equator, on those two icebreakers.

But first there were long transits from New York and Boston to reach the ice edge on *Northwind* and *Atka*.

Northwind and *Atka* were sister US Coast Guard cutters of advanced design but very different stories. As for design, each had one-and-five-eighths-inch welded steel, 64-foot beams and rounded hulls to allow rolling back and forth to break free of ice. A bow propeller would wash loose ice from in front of the vessel. At 270 feet, the ships were relatively short to facilitate maneuvering among ice floes. The hull forefoot sloped down from the bow, allowing the vessel to ride over thick ice and use its weight, 6,500 tons, to crack the floes, then back away.

Their design was as good for passengers and crew as for breaking ice. Riding an icebreaker in the Arctic pack ice is wonderfully different from riding the ocean waves, especially for those prone to mal de mer. With their wide beams and round bellies, these ships rolled farther but slower. They did not snap back like the research vessels I had ridden before.

Atka had the more interesting history. Launched in 1943 as Coast Guard Cutter *Southwind*, she saw a bit of military action near the end of WWII. With her sister ship, the *Eastwind*, *Southwind* captured a German weather station on Greenland, along with their supply trawler. My University of Wisconsin civilian scientist colleagues and I would occupy the same bunks as the captured German weathermen 22 years earlier.

On March 25, 1945, just two weeks before Nazi surrender, Southwind was transferred to the Soviet Navy under the Lend-Lease Program. The Soviets renamed the breaker the *Admiral Makarov*, honoring a pioneer Russian naval architect they consider the father of modern icebreakers. In 1950 the Soviets returned the vessel not to our Coast Guard but to the US Navy, which renamed her USS *Atka*. She was home to 14 officers and 205 crew.

My stint as a geophysics graduate student on the vessel from summer to fall in 1966 would be its last under the US Navy. Late that year she got her original name back, *Southwind*, and her Navy gray was repainted Coast Guard white.

In 1970, *Southwind* had made a courtesy Murmansk port stop in the country she had served for five years as *Admiral Makarov*. It was a reminder of her Soviet service times that were well known by her Navy crew and invariably pointed out to everyone newly on board. Somewhere in the engine room was a metal plate with Cyrillic lettering.

Southwind was decommissioned in 1974 and scrapped in 1976.

Northwind added a chapter to its and my history on the 1965 cruise, bending a prop upon first encountering pack ice, so we steamed back to Newcastle for repairs. That took more than a week, giving me as grad student time to travel to Edinburgh (saw a James Bond reenactment around the castle) and then to Fort William, from where I hiked up Ben Nevis.

The two missions were also unique, The 1965 *Northwind* cruise had a Cold War geopolitical goal: It was to be the first post-war Western (non-Soviet) ship to transit the Northeast Passage west to east from the Atlantic around Siberia to Seattle. Onboard journalist Richard Petrov documented the cruise in his non-fiction book, *Across the Top of Russia* (1968).

That mission brought us into conflict with Soviet mariners more than once. The issue of contention was the Vilkitsky Strait at the northernmost Siberian mainland. The US contended the Strait connects international waters as does the Strait of Gibraltar, so its width is irrelevant. However if the Vilkitsky Strait were 20 or more miles wide, it would qualify as international waters.

En route to the strait, we were tailed by various vessels and buzzed by Badgers, the Tupolev Tu-16 twin-engined jet strategic heavy bomber. The destroyer 020 came so close we could see Soviet sailors with binoculars and cameras lining the railings. *Northwind* contacted 020 and invited their skipper to come over for dinner. Long time no answer. As one of the only Russian speakers on board, Peter Vogt, the lowly civilian, would have been included in this dinner. I boned up in my Russian dictionary. I can't recall if the skipper of 020 sent his regrets. But the destroyer pulled far away and never came near or communicated again. Likely was that because Moscow higher ups feared defections.

In the Strait, *Northwind* used radar to show it was not quite 20 miles across. We anchored just west of the entrance— no pack ice in sight—with a Soviet missile frigate anchored nearby. The skipper, Capt. Kingdrel Ayres, and most on board including me, hoped we would proceed and call their bluff. But US State Department said no. Most everyone, especially the crew, were disappointed; they hoped for medals in Seattle.

Instead, *Northwind* tried to steam north around the Severnaya Zemlya (North Land) islands but was stopped by thick polar pack ice of the Laptev Sea. So we went back east into the Barents Sea and I got more data for my PhD thesis. *Northwind* had, after all, claimed this was part a research cruise.

Atka's special mission had highly personal significance for me. USS *Atka* would be met by my new fiancée, Randi Stampen, when the ship stopped in Bergen enroute to the Arctic, 1966.

Randi Stampen waiting for me on *Atka* on Bergen, Norway dock in August, 1966. We got engaged before we headed north. The crew teased and labeled me "14-day wonder" for getting engaged just after the Atlantic crossing from New York.

One of our missions involved not me nor my PhD thesis interest but my professor, Ned Ostenso, who was on board to drop explosives into the ocean and detonate them at a fixed depth. The propagation speed was to be determined. I think but don't know if the receiving hydrophones were suspended from Ice Island Arlis or T-3. These were tabular icebergs—rare in the Arctic—with manned stations. [These ice islands eventually got caught up in the ice exit flow out of the Fram Strait and were evacuated off Greenland before they broke apart and melted.]

Peter perched for a gag photo atop plastic explosives aboard icebreaker *Northwind* in 1965 He claims that his cigar was not lit.

While a packet of 50-pound Composition B explosives was strapped together and sitting on deck on a rare sunny day, I sat on that packet, pretending to smoke a cigar. In reality Composition B explosives require an electric detonator and can't be exploded by a match or cigar.

My professor had forgotten to bring floats from which to suspend the packet to make sure it exploded in the sound channel at the right depth. So he or someone resourceful hit on the idea of bundling together balloons, which many military ships carry for special occasions, like birthdays. This worked well. The detonation was by electronic remote. We heard it, and it caused a ripple on the ocean surface but not a fountain of water.

The Soviets tailing us far too close astern learned to back off when they spotted the balloon bundles. We had our fun with those ships. One time we just dropped off a bundle of balloons while they were close astern. Fun and games in the golden days of the Cold War.

The Soviets always fished out any of our floating trash. So one time the *Northwind* crew made a bag of *Playboy* magazines for the horny Soviet sailors to fish out of our wake.

My mission was the same on both voyages: investigating what lies below the Kara, Barents, Greenland, Norwegian, North, and Labrador seas and the Arctic Ocean.

One method, deploying a University of Wisconsin magnetometer sensor (colloquially maggie, maggot, or fish) behind the vessels, gave us 46,000 miles of magnetic data.

Much of the data mileage was collected on transits to and from ports or survey areas. We ran a few closer-spaced parallel tracks across the North Sea and across what in my thesis is called the Atka Ridge, the segment of the Mid-Oceanic Ridge running north in the Greenland Sea. Alas, the Soviets had already named it as Knipovich Ridge.

Over the underwater continental crust under the North, Kara, and Barents seas, magnetic field-strength anomalies reflected ancient, deeply buried igneous and metamorphic rocks. Over the deep ocean basins, we were measuring mostly strips of oceanic crust with alternately normal and reversely magnetized basaltic rock. Such anomalies had just recently been correctly interpreted in the 1962–1966 period.

A second method was to deploy two gravimeters to measure the small changes in the strength of the gravity field. Denser rock masses deep below the ship caused the force of gravity to be a bit stronger. It can be very challenging to measure on a moving and vibrating ship, and we only succeeded when the sea was calm and we were inside the pack ice.

In the Kara Sea, *Northwind* was stopped in the ice for 10 to 15 minutes with engines off. Setting up our meter on the after part of the main deck, near the ship's center of gravity, we got 24 good readings.

In the Barents Sea on *Atka* there was too much vessel motion, so we only got three good data readings, those when it was calm and all engines were down for repairs. The other 30 good measurements we got while on the ice. Of those, we made four by being helicoptered 4 to 10 miles from *Atka*. We chose large ice floes, and the chopper hovered around while we leveled and read the meter. We discovered that metal folding chairs with concave seats provided a good base on which to level the meter.

Twenty-six more gravity measurements were made while inside the pack ice in the northeastern Barents Sea. The icebreaker would stop, secure engines, and two or just one of us would climb down the Jacob's Ladder onto the ice and set up the meter.

Such maneuvers were not always carefree. On one occasion *Atka* began to move, opening up a seawater gap between the hull and the ice floe. It happened fast, and the bottom of the ladder was still hung up on the ice. I abandoned the meter and jumped on the ladder, which then detached from the ice and sent me banging into the hull.

The icebreaker did not move much more, and a Navy seaman climbed down on the ice from the bow later and managed to retrieve the expensive instrument.

Because Earth is a bit wider at the equator, we weigh more near the poles. This effect has to be subtracted from our readings. So we needed good navigation. On *Northwind*, we got a few early satellite fixes, but navigation in the mid 1960s still depended on the stars—and in the Arctic summer on the sun. This was problematic because of the nearly constant stratus overcast. Radar fixes from the low-lying islands were also poor. Ushakov Island was 30 miles from where the Soviet chart had placed it.

The third bit of planned geophysics was to deploy seismic refraction to measure sound velocity structure below the seafloor. We shot only one line in 1965 and one in 1966, both in the eastern Barents Sea west of Novaya Zemlya, one of two major Soviet nuclear test sites. The procedure involved men (not me) dropping Nitramon explosive charges of 0.5 to 50 pounds from the ship's Greenland cruiser along a line from 0.5 kilometers to 10 kilometers from the ship. The explosives were detonated by radio, and the sound waves traveling through the water and also through the upper sediments were recorded by hydrophones suspended from the ship.

Most onboard technicians were not from the University of Wisconsin but from the US Naval Oceanographic Office, where I would work just a couple of years later. While collecting water samples, they took measurements of water depths and water temperature.

As I have said, most of my geophysical data was collected while steaming from one oceanographic station to the next. Approaching icepacks was a story of its own.

As we approached, a predicted pack ice edge would show up clear on the radar, especially if the wind was blowing toward the pack, cramming the floes together. If the wind was blowing off the pack, some ice would be blown farther than others. In that case we first saw isolated bits of ice.

Sea-ice people describe the pack in units of 10 percent ice cover. If the sea is completely covered it's 10/10; 5/10 means half covered. The gaps between large floes are called polynyas, a Russian word.

We were in the pack during the late summer, the time of minimum ice cover. Summers in the Arctic Ocean are commonly overcast with low stratus cloud decks. While they hide the sun, such clouds can tell pack-ice mariners if open water lies ahead. It's called water sky. The underside of the low cloud deck reflects the dark shade of open water.

Today, as on the 1965–'66 cruises, icebreakers also carry helicopters able to scout out the pack ice ahead and look for wider polynyas.

Most of the summer ice was yearling, surviving from the previous winter: relatively thin (2-to-4 feet thick) and, unlike freshwater ice, internally a mix of ice and brine. Yearling ice becomes frozen solid only during the polar night at temperatures far below zero degrees Fahrenheit. With summer air temperatures around 30 to 35, most floes were covered with some wet snow, slush, and even puddles.

Some ice floes were small, castle-like remnants of pressure ridges from winds and currents that pushed floes together to form analogues of mountain ranges like the Himalayas and Alps, where tectonic plates collide.

Both breakers simply plowed through the rotten yearling ice. Anywhere near the hull, we heard a combination of swishing and scraping. Now and then this was replaced by jolts and bumps, much like riding on a truck over bad roads with potholes. Then we knew we were traveling through older ice.

Ice floes repeatedly melting and refreezing over some years gradually expel their brine and become more solid and thicker. Standing near the bow, I saw such floes being overturned and exposing beautiful powder-blue interiors. Older, multiyear ice thickens to more than seven feet and much thicker in pressure ridges. At some point, icebreakers have to resort to riding onto floes to crack them, then backing off and repeating.

We saw similar powder-blue colors but with layering when, rarely, we encountered chunks of glacier ice. Today there are very few glaciers flowing into the Arctic Ocean. The icebergs of Iceberg Alley, like the one that sank the *Titanic*, calve from central Greenland glaciers and are carried south by the East Greenland current. The same current carries unlucky Arctic Ocean multiyear ice floes out of the Arctic Ocean.

Now and then, we saw polar bear tracks—broad pawprints of animals evolved for swimming and distributing their great weights over new thin ice to avoid breaking through. Swimming prowess gave this bear its scientific name, Ursus maritimus.

Their Russian name translates to white bear, while the German and Scandinavian names translate to ice bear. The English name Arctic Ocean also relates to bears, deriving from the ancient Greek word for bear as in Ursa Major (our Big Dipper) constellation with its Pole Star. The German and Scandinavian name for that ocean simply translates to Ice Sea. The Russians call it the Great Icy Sea.

As a species polar bears are, as Homo sapiens, young, having evolved from brown bears and occasionally interbreeding with them in Siberia. Today the ice bear is the charismatic poster child for climate warming. We saw one or two on both the *Northwind* and *Atka* expeditions. I recall that the *Northwind* skipper wanted to bag one from the ship. I hope I misremember.

USS *Atka* Ice Liberty, 1966. After a long time on the icebreaker,
ice liberty was heaven for these Navy sailors.

Walking around on 5-to-10-foot-thick sea ice during so-called
ice liberty was a strange experience for me. I knew that I was
standing still or moving slowly above an ocean thousands of
feet deep. But I was feeling as if on some Great Plains farmer's
bare snow-covered field. Except in the Kara Sea, our gravity
meter could always detect motion from long-period swells
from the open ocean far away.

This experience was only trumped once for me—in the
tropical South Atlantic, aboard the oceanic research ship
Robert Conrad, in 1986. The vessel was stopped underway for
a swim call. A crew member with a rifle stood watch up high in
case of sharks. I swam in a clear, warm, deep-blue "wine dark"
ocean probably two miles deep. Swim calls were later abolished
by *Conrad's* operator, the Lamont Doherty Earth Observatory
of Columbia University.

However I believe the US Coast Guard still offers ice liberty,
but that will also someday end—when the Arctic Ocean
becomes ice free in summer.

Atka finished its northern Barents Sea deployment in early October, 1966, late enough to let me experience the first stage of a freeze up. The air was now below freezing, so tiny ice crystals called frazil ice were forming in the ocean. A thin layer of such crystals coated the ocean surface, preventing capillary waves from forming. The sea surface looked like it was covered by an oil slick but without colors. This is called grease ice. (I have seen frazil and grease ice in the Chesapeake Bay during very cold winters.) The next stage comprises small round pancakes with upturned edges from bumping into each other.

The pack ice was its most beautiful when the sun came out and there was no wind. The glassy open water reflected diversely sized and shaped floes. Still as a mill pond it was, but deadly cold. We were told that falling into 28-degree Fahrenheit water from the ship spelled death. The icebreaker could not turn around to rescue us in time.

The USS *Atka*, which back in Bergen had been met by a very special person when we returned from our mission in the Barents Sea.

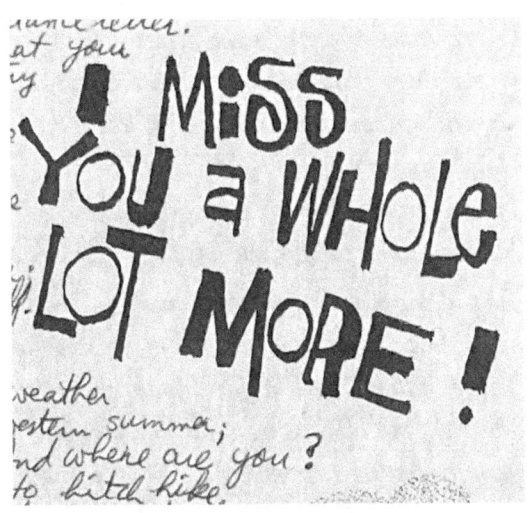

Letters Home

Writing about events more than half a century later takes a good memory or surviving contemporary notes. The journal I kept on Northwind *survives, returned to me after Richard Petrov used it in writing his non-fiction book,* Across the Top of Russia *(1968). For the 1966 Atka expedition, letters were sent back and saved. The 1965 and 1966 icebreaker cruises collected data for my PhD thesis, which fills in dates and measurements but obviously omits personal experiences.*

During pre-cruise summer 1966 I was working up 1965 Northwind *data but suffering from mononucleosis. The girl I had been going steady with was in Norway visiting her relatives and ancestral acres, so I wrote to her there from Madison and while* Atka *was still in port from the ship; at least then, Navy vessels carried a postmaster. The stamps I pasted on the envelope were canceled by USS* Atka *on 28 July 1966, the day before ETD.*

Two days out a crew member got acute appendicitis and had to be medivaced to Newfoundland, so I mailed another letter to Norway from USS Atka *on July 30. How did I mail those three icebreaker letters?* Atka *made no port stops while in the Arctic. I must have mailed them when back stateside.*

Some may think I dropped them off on a mail buoy, but this is a traditional prank played on newbies on ships, perhaps since the late 19th century. The newbie is told that the Post Office collaborates with the Coast Guard to set out mail buoys where Navy ships can later deposit or collect mail. The hapless newbie is supposed to keep watch for the buoy and snag it with a hooked pole.

Mail buoy watches are akin to snipe hunting and watching for the Line (equator) by pollywogs on Neptune's Day. Helps relieve boredom on long ocean voyages.

These letters were to my new fiancée, and personal comments have been omitted. I wrote the first three letters from the East Coast before we crossed the Atlantic to Bergen, Norway.

The descriptions of pack ice, weather, polar bear tracks, and buzzing by Soviet Badger aircraft, as recorded in letters to Randi from the Atka *cruise, apply also to the* Northwind. *cruise. I also wrote letters from* Northwind, *but they went to my parents and burned with their house in 1977.*

Undated: From the Hotel Essex in Boston

Dear Randi,

Don Nelson (our tech) and I had a cool drive out—26 hours on turnpikes. The Mass Tpke through the Berkshire Hills on a sunny day— smoking a cigar (my tonsils are well enough that I harass them again), barreling down the turnpike (This truck was owned by the Geophysical and Polar Research Center of the University of Wisconsin with $250,000 in electronics aboard—today $2.5 million) and not a care in the world on a sunny afternoon in July.

Drove down to the *Atka* this morning, met a few officers, arranged a magnetometer to be spot welded to the bulkhead, etc., had lunch in the wardroom. Most personnel are still on leave. Some of the officers seemed quite bleary eyed and hungover from excessive night-living. Ran some errands today, including picking up a mag sensing head at the airport. Boston is a wild and woolly place to drive around on an errand.

The hotel is only a mile from the *Atka*, behind which, at the next pier, looms the great grey hulk of the USS *Wasp*, the carrier that picked up some astronauts who fell into the sea...

In a few days I will be lurching around the gray decks and clammy lockers and bunk rails...

It's 11:30 in the hotel room; I have a snort or two of blackberry brandy I took along, and a radio plays "Greensleeves," and two gravity meters sitting faithfully on the dresser, their thermostats clicking occasionally like contented hens...

28 July, 1966
0945

Dear Randi,

In wardoom of *Atka* at sea a few miles off Long Island...presently approaching Earle, NJ to take on explosives & ammo.

Pulled out of Boston Navy Yard under sunny skies; many heart rending good-byes on the pier. One unfortunate sailor, the ship's photographer, had both his wife and his girlfriend on hand; he deftly maneuvered out of the situation, but last seen on the disappearing pier, wife and girlfriend happened to be standing next to each other...

Whales sighted yesterday, a common sight around the Atlantic Wall.

Will begin towing our magnetometer tonight or tomorrow...

Undated

Dear Randi,

We are progressing toward Bergen with yet another opportunity to drop mail off in North America. This letter will accompany an officer with acute appendicitis, being flown from 20 miles out in the cold green Labrador Current, via helicopter, to Halifax, Nova Scotia. Bright sunshine alternates with surface fog. Last night the stars above were bright but you couldn't see from bow to stern. Every minute the hoarse bellow of the foghorn explored the gloom...

Life is dull as yet; we are waiting until off Cape Race to get the "maggot" into operation, the latter term of endearment being a specialty of our fresh young technician.

Have begun the monumental chore of studying for prelims.

20 August 1966.
Steaming east on 75th parallel,
200 miles east of Bear Island, midnight

Dear Randi,

I had to prepare for my PhD prelims coming up in mid November. (Prelims are oral exams with my entire committee present to fire questions across a wide range of geology and geophysics.)

Seas continued flat or only moderately choppy as we steamed north into the icy waters of the East Greenland Current. An ocean station: Water column sampling and temperature measurements were done by other programs so I just observed.

An ocean station on 16 August at 75° N 100 miles east of Greenland began the activities. The next day the oceanographers lost 2000 meters cable and a string of six pairs of reversing thermometers (loss $2500) and Nansen bottles to Davy Jones. I was standing around idly when SNAP! the cable broke. TS. So while they diddled around putting on a new cable, we prepared for our first seismic line, scheduled for the next day.

A trial shot and recording went off ok. All systems go, and the Greenland cruiser was readied.

However, after several weeks of calm weather, when do the seas deteriorate? You guessed it right. The captain called the launching off the next morning. Murphy's Law...

The last three days have been rather rough, except on legs where we have following seas. Who knows when the next opportunity for a seismic line will come. As far as my interests go, we missed the best opportunity, i.e., to determine whether and how deep there is Mid-Atlantic Ridge flank topography buried under the Spitsbergen continental rise. I did manage to finagle a couple of east-west runs across the rift valley, substantiating our last year's [*Northwind* 1965] finding that, unlike elsewhere, no magnetic anomaly is associated with it. The echo soundings show a series of step-like drops into the valley, each drop, possibly a normal fault scarp, amounting to several tens of fathoms...Bad weather has also ruined an ideal opportunity to fly [*Atka* carried two helos] to nearby islands, Bear Is. & southern Spitsbergen, for gravity stations.

The pilots are getting intensely bored with nothing to do except chow, movies, card playing and rack time...

No ice yet, no sign of Russians. Grey sky, grey seas, grey ship, decks wet with spray and sleet and rain. But spirits are generally good, meals in the wardroom being the high points of the day. In deference to exams, I have given up movies.

Naturally for the first few days out I (and you) [I had gotten engaged to my future wife Randi during our Bergen port stop] were the talk of the ship. Even the galley cooks carrying garbage down the gangplank—well they weren't THAT intent on their work not to notice us. And, John Andrew being the Hedda Hopper he is, everyone knows we are engaged. I have had to put up with much ribbing & advice (e.g., 'the 14 day wonder,' 'you fool').

30 August 1966
In Latitude 78° N

Dear Randi,

Outside the porthole is a gray-white world of pressure ridges, dark polynyas, blue-green meltwater pools already covered with an inch of new ice. The air is freezing, the sky a monotonous overcast, except for dark patches of 'water sky' above the polynyas. Slowly the *Atka* grinds and crunches and lurches northward, curiously, in search of open water. There are flurries of snow and sleet in the damp cold air...

The big pads of polar bear tracks meander about the floes now and then, new tracks left in the inch or two of new snow. It is just about breakfast time, a meal I have not been in the habit of waiting up for.

The last 48 hours have been wearying, only an hour sleep here and an hour there. We are making hay, etc, by reading our gravity meters while in the ice; we couldn't get even fair readings on board, so it's over the side, John Andrew and I, on a rotten rope ladder, onto the sea ice below. Some of the ODs have us tied to safety lines as we climb down, lower the meter cases carefully, and hunch over the little bastards, as some ritual or prayer, leveling, turning those spring tension dials, trying to average out the swings and bounces due to the imperceptible heavings of the ice and noise from the ship, whose gray hulk looms above us, a maternal monster. Six times we have gone over the side the last 24 hrs, about every 10 to 15 miles.

Yesterday the ship was actually stuck in the ice for 12 hrs. Stuck fast! We backed and backed; swirling boils of ice-choked water aft, to no avail. Heeling tanks didn't help either.

Finally the ship was free, due to hoisting the 15-ton Greenland Cruiser (round-bottom metal-covered boat on main deck) over with the crane, getting a lever arm to tip the ship over & free it.

The Russian ship, with only 4000 HP, decided to chicken out at that point, after tailing us for days. Now it's the Badgers again [western name for a Soviet aircraft], with one dropping two red flares this morning.

1900 same day

Bleary-eyed I climbed out of the 'tree' for chow. Miraculously the view is a gray sea, nary a piece of ice. The five days in the ice seem like a dream. The *Atka* plows westward along the 79th parallel, laying ocean stations every 30 miles.

This morning, after I began this letter, there was a call for flight quarters, and John and I lifted off the helo deck, decked in fancy, ear-pinching crash helmets, inflatable life vests outfitted with whistle & flares and such goodies. The gravity meters we held between our feet.

The helo flits above the pack at 50 kts, perhaps 1000 feet up. We had wanted to read the meters at 5 and 10 miles abeam of the ship's track, but the grey sky on the horizon, water and sky indicated dark water below. The pilot pointed out a likely, football field-sized floe below, wedged among smaller ice...

6 Sept 1966

Dear Randi,

For the last 48 hours the *Atka* has lain motionless in wet foggy pack ice in 78° 30' N, northwest of Novaya Zemlya.

The reason: A short circuit and fire in a main motor control panel. It was timed by the gods to happen on the 4th, Sunday.

Monday was Labor Day holiday routine, and it happened! After several hours of cautious maneuvering on the remaining shaft, *Atka* backed into a huge flow, an ice anchor was put out, and beer went on sale: 12.5 cents per can! A mass exodus over the Jacob's Ladder. There was soccer in the snow (one ball went immediately into the icy water), 'king of the pressure ridge,' sledding down a 15-foot pressure ridge and, naturally, some brawling. Ensign Nelson (Don Nelson, our University of Wisconsin tech) disgraced the State of Wisconsin by getting sick after a mere seven cans of Heineken. It was about 34 F, light snow, and fog.

Nanook, who has been howling and whining every time one of us goes over for a gravity reading, had a field day. Of course the dog had to be carried up & down the Jacob's Ladder in someone's arms.

The flow is deserted now, save for a field of beer cans, a chair, and a gravity meter. An hour ago, after the ship had lain motionless for hours after the ice anchor had been pulled out, I decided to read the meter one last time (dedication!). We have a smooth curve of readings since we came to a stop 48 hours ago, the variation being due to our drifting with the pack, at an unknown rate, in an unknown direction (due to the fog: no celestial fixes).

Anyways, no sooner than I had untied the meter, CRINKLE! CRUSH! The ship began to edge away from the floe. I thought it was under power, so sent Nelson running to the bridge. Meanwhile the gap was widening (suspense!). I untied the meter and stood there, with typical Vogt-type indecision. The end of the ladder was caught on the ice. Science or security first? Yeah! I grabbed the ladder, which swung free, and donked into the ship, and climbed up.

PART III

A Navy Scientist Explores the Depths

A ROAD LESS TAKEN: NAVAL RESEARCH

In 1967, I was offered a job in DC with the US Naval Oceanographic Office. The Office (NAVOCEANO or NAVO) was then a mapping and charting institution that served the US Navy. NAVO sought practical information: seafloor topography; navigation hazards; ocean currents; and the ever-changing deviation of the compass between magnetic north and true north.

All those practical matters are also topics for scientific research. Dr. Charles C. Bates, popularly known as "Charging Charlie," was trying to change NAVO culture by adding research as a mission. That's why he came to the University of Wisconsin and recruited me, at the recommendation of my major professor, Ned A. Ostenso.

I had a Master's in Oceanography and was about to earn a PhD. I already had ocean-going science experience, including on Coast Guard (1965) and on Navy (1966) icebreakers. At the time I was considering a university position in northern Arizona and was not keen to move to the hot, humid, over-crowded Mid-Atlantic, devoid of mountains and cold winters.

Ostenso convinced me to take the NAVO job. He said I could stay there for two years just to learn how things work in the US capital. I ended up spending not just two years but essentially the rest of my life in the area.

In the August, 1967, DC heat, I started work in the Earth Physics Branch.

At NAVO, I had very little contact with higher level management like Bates. (He left some years later for the US Coast Guard. Only many years later—both Bates and I were retired, in his case in Phoenix—we got back in email contact. He consulted me for his 2006 book *HYDRO to NAVOCEANO*.)

Most of my comrades were good men. (No women then, nor Blacks, Asians, or Hispanics worked in our branch. In fairness, however, NAVO as a whole had already then a much more diverse workforce than any private concern.) But they were scarcely university-grade physicists. I was one of few NAVO civilians with a PhD.

Privileges of the PhD

I could get quicker results from the NAVO bureaucracy by identifying myself, on the phone or on paper, as Dr. Vogt. By contrast, when I later moved to the Naval Research Laboratory, NRL, let alone when I occasionally gave seminar talks at universities, PhDs were a dime a dozen, so we rarely mentioned our degrees.

Office dress was another form of employee expression. At NAVO most of us dressed semi formally. Later at NRL, we scientists down-dressed, irritating the uniformed naval officers. Only the Xerox repair folks and beltway bandits (contractors) wore suits and often ties.

In our branch, only one man, branch head Thomas Davis, was away working on his degree at Penn State. Davis later became a kind of star at NAVO, thanks to his applied-math work in time-series analysis, making sense of wiggly line data. (Although his unclassified analysis techniques helped the US Department of Defense, it did not—as Tom once confided—help him get rich predicting the stock market.)

The Cold War had led to many new hires, including those not educated in fields relevant to the needs of the Department of Defense. When I arrived in DC, there was obvious dead wood, and firing someone for incompetence was all but impossible.

I had been working in the Earth Physics Branch for less than a year, commuting from Oxon Hill in Prince George's County, when another man came to NAVO to promote research there. A red-haired lieutenant, Eric Schneider, was about to get his PhD in marine geology from Columbia University, which includes the off-campus Lamont-Doherty Geological Observatory. He was brash and ambitious. In his book, Bates referred to him as "the wild man."

My wife, Randi, and I got a whiff of the wild man at a party hosted by a branch member and his wife. Party conversation was abuzz with alarm. It was about Schneider,

Schneider held two face cards at NAVO and played them both. He was a naval officer and a PhD. This trumped the civilians—including NAVOs middle-aged senior managers.

Schneider's game plan was to bring down from Columbia the research and technical talent he could, hire others as needed, and, unlike Bates' plan, start a brand new and separate research group at NAVO. In this innovation, he was enabled by the NAVO Commanding Officer, Capt. Tex Treadwell, and by his major professor, Bruce Heezen, who had been and was then still an advisor to NAVO. The Navy wanted an academic researcher to help oversee and vet NAVO programs, presumably at least the unclassified ones.

Both Bates and Schneider wanted to promote credible basic research even inside NAVO. By credible, I mean publishable in major peer-viewed science journals.

Peter Vogt getting an award for a publication.

In Navy terminology, a 6.x system is used to label research. Basic research with no conceivable future naval relevance would be 6.0. Research done at NAVO with results published in science journals, which might be about sub-marine volcanism, would be 6.1. Of course NAVO would also conduct classified research and maintain classified databases generally not available to academic researchers or foreign entities. For example, chemical coatings making torpedoes travel faster might be, say 6.4+, and classified.

Human judgment is involved in deciding what data should be classified and at what level. Back then but since abandoned, the lowest level was called Restricted. Then Confidential, Secret, and Top Secret. Above that there is compartmentalized Top Secret.

Neither I nor any one of my Naval research colleagues even considered taking any classified documents home. In fact we took our turns checking that all classified documents checked out for office inspection had been returned to our safes, and the safe or safes locked and initialed before we left for home. The idea that some future day an ex-president would hold highly classified documents and refuse to return them to the US government would never have occurred to anyone.

Schneider's new NAVO group would be called GOFAR (Global Ocean Floor Analysis and Research). Technology could have been added, but for obvious acronym reasons was not. One day the brash young (my age!) Lieutenant PhD called me to have lunch at an Italian restaurant in Suitland. I brought along an Earth Physics Branch colleague, Angelo Kontis, and the two of us listened as Schneider explained what GOFAR was all about and that from what he'd heard, I would be a good fit. That would make me an exception; I was already working for NAVO.

Although both Anj and I later agreed that Schneider was too bombastic and full of himself, he was a spirited leader and a good scientist. I didn't have much pity for some of the elderly NAVO deadwood managers Schneider unfrocked.

My PhD thesis advisor at the University of Wisconsin, Ned Ostenso, opined that GOFAR was "playing tennis on a football field," the latter meaning main NAVO. At some point Ned left academia to head the NOAA SeaGrant program.

The new GOFAR group did not yet have its own office, but most were housed in the Washington Navy Yard. The technical illustrator Schneider hired, Barbara Grosvenor, lived with her laser physicist husband Albert on base at the Randle Cliffs NRL facility on the top of the Calvert Cliffs just south of Chesapeake Beach. This facility had been built there in WWII to test radars on target ships out on the Chesapeake.

Grosvenor reported that NRL had ample excess space there, including an unappealing empty WWII cafeteria. A large block house building was already being used by the NAVO Magnetics Division. So GOFAR (about three dozen employees) was established there, and many of us moved to live in Calvert County. Some of our secretarial staff was hired from the county. Schneider was pleased with this site, 45 minutes from NAVO headquarters, because it gave him and his new group some sense of autonomy and independence. However, this ran counter to military and other management protocol. Their eyebrows rose further when Schneider produced a GOFAR promotion blurb referring to it as "an institute."

The Schneiders now owned the historic English colonial Patuxent Manor in Lower Marlboro, Calvert County, Maryland. The Duponts had long ago stripped it of interior paneling, but the brick exterior was authentic. One GOFAR oceanographer with artistic aspirations, Ed Escowitz, soon set up his easel in the neighboring corn field and began to paint the home. A few years later the Schneiders sold Patuxent

Manor and acquired an even more famous historic home, the cruciform-shaped Cedar Hill. While the Schneiders owned Patuxent Manor, its historic character and spacious backyard made perfect for the 1968 Kane-9 bon voyage party, a party mainly for GOFAR families.

GOFAR's first and much-heralded research cruise was the Kane-9 expedition to trace fracture zones across the central Atlantic in summer of 1968. I would have been onboard. However I used my new professional ties with the Magnetics Division to go as watchstander and scientist on that seven-week Project Magnet circumnavigation of the southern hemisphere in early 1968.

CHAPTER 1

Kane-9 Insurrection

I can't recall if Bruce Heezen, chief scientist on Kane-9 (and Schneider's mentor) was present at the bon voyage party.

Heezen's goal on Kane-9 was to identify and trace several of the fracture zones that cross the Mid-Atlantic Ridge and at the ridge axis offset it. The most prominent of the fracture zones traced on Kane-9 has long been known as the Kane Fracture Zone.

Mapping the bottom topography of the central North Atlantic the Mid-Atlantic Ridge was a scientific objective of the voyage. The mission also involved taking deep-sea sediment cores, dredging rock, and photographing the seafloor.

Below Heezen's area of study, hot magma would rise and spill out on the seafloor and the crust occasionally crack, producing earthquakes. This was the boundary between the North America and Africa plates. Today, geology recognizes that the Africa plate is actually two plates, the Somali and Nubia plates, the latter one extending west to the Mid-Atlantic Ridge axis.

The fracture zones to be traced on Kane-9 were also marked by mile-deep valleys and in places bordered by mile-high ridges. However there was no modern seismicity or volcanism along these fractures—except where they crossed the Mid-Atlantic Ridge.

Before the voyage began, the problem of illustrator and cartographer Marie Tharp being onboard this naval ship had to be solved. Until then, no females were allowed, allegedly for their own protection from numerous horny sailors. The Navy agreed to take Tharp on Kane-9 only if she shared her cabin with another female. Lois Means, a GOFAR secretary, volunteered to be that female. As it happened, she later married a sailor she had met on that cruise.

I did not follow news radioed from USNS *Kane* while she was at sea. However, I heard bad news upon *Kane's* return. Nothing technical, nothing geological. It appeared that chief scientist Heezen and maybe also cartographer Tharp had scandalously misbehaved.

Nothing salacious! Heezen had insisted the shipboard lab record "stomach time" so he could later remember whether a fracture zone was crossed before or after dinner. Unheard of and confusing! The long-standing and research practice on ships at sea is to record only Zulu time (GMT). Defending their revolutionary and conservative positions, Heezen and Tharp had shouting matches up in the chart room, throwing chalk and erasers at each other. There was more that I don't recall a half century later. These folks were senior professionals, representing Columbia University and the US Naval Oceanographic Office.

After the *Kane* docked, Heezen took all the seafloor photos, loaded them into his trunk and drove back north to Columbia. These photos (taken by GOFAR's Walter Jahn) were NAVOCEANO (US Navy) property.

In the weeks that followed, GOFAR chief Schneider led an insurrection against his former prof. The photos had to be returned to GOFAR forthwith. Heezen should apologize for his childish behavior at sea and be removed from being a NAVO academic advisor. Schneider and

Heezen accused each other of exploitation and dumping one another once no longer useful.

The NAVO commanding officer, Capt. Tex Treadwell, was forced to deal with an internal insurrection. He called a special meeting, a kind of hearing, to hear both sides. Although physically a small man, Treadwell appeared large behind his large desk. I was invited as an honest broker because of not having participated on the Kane-9 expedition. Cruise participants recounted the misbehavior stories. I mostly sat and listened. Heezen sat in silence, I think. Tharp was in tears. I don't know what Treadwell later decided, but Heezen never visited GOFAR again. We all went back to our work.

Capt. Treadwell led NAVO to the organization's peak, according to "Charging Charlie" Bates, the man behind NAVO's addition of a research mission. After Treadwell retired, he became oceanography prof at Texas A&M. We at GOFAR presented him with a coffee table paved with seafloor rocks sawn into slabs.

In 1977, nine years after Kane-9, Bruce Heezen died of a heart attack aboard nuclear research submarine NR-1, a sub I would ride 1999, and a topic later in this book. Both Tharp and Heezen had Navy survey vessels named after them, great albeit posthumous honors.

CHAPTER 2

Classification Conundrum

As the only GOFAR member with professional ties to the Magnetics Division, I regularly walked over there to marvel at some of their data, especially the classified maps showing from the pattern of stripelike anomalies where the northern North Atlantic had opened by plate tectonics. The division staff spent months at sea and more months processing the data.

However, analyzing it geologically and publishing the data was not part of their mission. They had no PhDs but some college background and knew far more about proton precession magnetometers, navigation, and data processing. Some had definite research interests, so in my position I was encouraged to coach/advise them, including giving them a chance to be first or co-authors on papers to be given at the American Geophysical Union annual meetings and later published in professional journals. My NAVO magnetics co-authors included Dewey (Ray) Bracey, Otis Avery, Charles Anderson, Robert Higgs, and maybe others. Bob Higgs became Division head and later led the entire Hydrographic Surveys Department.

Even before GOFAR came along, Higgs had recognized the new value of magnetic data over the oceans and had encouraged staff to research where possible. From Cambridge University research, he recognized NAVO magnetic data from the South Pacific supported seafloor spreading. But upon submitting his unclassified conclusions in an in-house technical report draft, he was ordered to cut that part out and leave geoscience to the academic community. Higgs had sent his unclassified magnetic data to the UK: Cambridge University researchers Prof. Drummond Matthews and his PhD student Fred Vine.

My NAVO Magnetics Division colleagues had always been allowed and encouraged to attend American Geophysical Union conferences and other such gatherings held back then at the Shoreham Hotel in Washington. They (I should say we) would listen and watch in silence as academics showed 35-mm slides with very limited data, and thus came to BS conclusions. Ironically, the Office of Naval Research was funding these university researchers.

Our oil company geophysicist cousins are often in a similar predicament. They have access to classified company seismic reflection data showing faults and other structures in great detail. These scientists also sit in silence and mutter BS under their breath at academic researchers with scant and/or poor data and consequently dubious interpretations.

The classification/censorship issue is complex. Of course I don't know much about oil company geophysics data, except that such contractor data is often held by a number of companies, all of whom have to sign off on any data release. Seismic reflection surveys along Southern Maryland highways were collected in the 1980s but have not been released even today. These data have great educational value, while their potential for revealing commercial hydrocarbons, as far as I can tell from reliable hearsay, is likely minuscule.

In my opinion, not all classified data, government or commercial, needs to be classified. However, declassifying data, unless automatically by its aging out, is not that simple. The US government folks who have to sign off on declassification may not know the value of the data to science, especially to American science. There are always gray areas. Bureaucrats get no benefit from releasing data to the public but only blame if those data later turn out to have value to a potential enemy.

In many cases sensitive data—military or commercial—can be sanitized by smoothing away details while retaining and releasing the low-resolution big picture. High-resolution images of the ground from orbiting spy satellites could be digitally blurred.

In the early years of GPS, the navigation accuracy was deliberately reduced by digital 'dithering.' For authorized users such as our 1984 Navy Research Lab aero-gravity work out of Ascension Island (more on that later), a classified 'p-code' was provided to get the high accuracy. Most of these DOD efforts were later nullified by legal commercial, international and academic prowess. The Ukraine-Russia war has been fought with the help of GPS-enabled drones, and most modern car drivers and smart photographers know exactly where they are, thanks to GPS.

In the case of oceanic survey data, the ship track locations (i.e. the navigation) are more important to classify than the actual data. In 1972 a GOFAR colleague (G.L. Johson) and I submitted a paper about Mid-Atlantic Ridge topography to the prestigious Bulletin of the Geological Society of America. The bathymetric map of a large area that had been sounded by NAVO survey ships had been sanitized to remove, by smoothing, fine details. But it was a vast improvement over what had been available.

The two peer reviewers liked the paper, but the journal editor, Prof. Tjeerd van Andel, took serious issue with it because we did not show the ship tracks on which the map was based. Good science, he said, must include metadata: how and where the data were collected and estimates of error margins. Van Andel allowed publication of our paper, but—perhaps uniquely in science literature—appended his objections, and allowed us to respond. In my rebuttal, I pointed out that geologists had generally not objected to oil company data being published sanitized and without track control. Moreover, any random well-navigated academic track through our map area could prove its accuracy.

What did my academic colleagues (and competitors) think of folks like me? Some seemed to think the government, the Office of Naval Research, and the National Science Foundation were there just to fund academic research, not to conduct their own, as we did.

Historically, the Naval Oceanographic office and its predecessors date from the middle of the 19th century. Many US university researchers and their grad students depend on soft-money grants to supplement what they earn from teaching. Research requiring ship time also requires government funding, with daily cost of a modern vessel being in the $50,000 ballpark. It's a lot cheaper to pay a researcher to write computer code and model processes, using existing data.

Some of my colleagues thought that people like me had unfair advantages looking at data denied to academia, then publishing what was unclassified. In fact only a small percentage of my (et al., et ux.) 150+ peer-reviewed publications was based on such survey data. Some of my most original and frequently cited publications involved my reinterpretation of data collected by my academic colleagues.

Grounded with one foot in academia but the other in a military lab, I had an unusual career. The Navy benefited because I kept up with all the advances in academic research and passed on discoveries of possible naval relevance. Yet I missed out on the university student world. The 14 postdocs I had at the Naval Research Lab over the years weren't really students. They knew more in many areas than I did. We ended up hiring some of them.

The basic research (called 6.1) we did at the Lab had to pass critical review by the NRL multidisciplinary Research Advisory Committee. We prepared well and somewhat feared the reviews. In later years our program was reviewed every three years by a mixed external academic committee, the same type of review our academic colleagues had to survive. This review system was imposed by the Office of Naval Research; I was mostly the branch scientist in charge of our reviews.

CHAPTER 3

Project Magnet— Zig-Zagging Over the Southern Hemisphere's Night Skies

At the Naval Oceanographic Office in early 1968, I signed up as science watch-stander on one of the round-the-world flights of Project Magnet—thus missing the Kane-9 expedition. Project Magnet was, in short, a major survey effort from 1951 through 1994 conceived to model the world's geomagnetic field (what turns the compass needle). Having published marine magnetic data and written modeling computer programs, I was welcome to participate.

We made the typically long flights aboard Project MAGNET's Super Constellation, *Paisano Dos.* I think Super Constellations are the most beautiful airplanes ever built—though this particular one sported a roadrunner cartoon character painted on the fuselage by an earlier aircrew. It also had a functional peculiarity: a non-magnetic tail section. The magnetometer was installed aft and so had to be as far as possible from artificial sources of iron.

The Lockheed Constellation dates from 1958, but *Paisano Dos* (Lockheed NC-121K) was specially built for Project MAGNET in 1962. The original *Paisano* had crashed, all surviving, in 1960, landing on the ice near McMurdo (where I would land and depart two years later). This amazing aircraft was retired in 1973 and finally scrapped in 1976. However, you can buy and assemble your own Super Connie model.

On our deployment in Naval Oceanographic Office Project A31, the flight schedule called for departing DC on January 9, 1968 at for its first leg to Paramaribo, Surinam. Besides collection of airborne geomagnetic data (strength and direction of the geomagnetic field), our mission at every airport we landed was to make land gravity measurements and monitor the reception of satellite navigation signals. We used not our modern GPS but an earlier method using Transit satellites.

Our schedule called for many stops at commercial airport facilities so advance clearances had to be obtained from many embassies. Moreover, reservations had to be made in advance at military lodging and hotels for civilians. We flew at night, arrived the next morning, and left again in the evening after two nights' sleep. The policy requirement for two successive nights rest is for the sake of the pilots. But we scientists (most with a good bit of tourist in us) did not mind.

Why fly at night? Back then, we depended on the stars for navigation. Plus, the atmosphere, especially in tropical regions, is calmer at night.

Project MAGNET was also a good way to spread good will, especially in areas suspicious of the US military. To make sure we were not US spooks, we had no aircraft cameras, only personal cameras to be used according to local restrictions and traditions. The compass was still widely used around the world and our data would update existing maps—available to all—showing time changes in deviation of the compass (the angle between true north and magnetic north). The data we collected would be available to all, and *Paisano Dos* would welcome public visitors at airports.

Paisano Dos carried 20 personnel, four or five scientists, five officers and nine enlisted. Our chief pilot was CDR Charles Hickey; the copilot, Lt. CDR Lou Lomheim. The aircraft carried up to 6,500 gallons of 115/140 octane aviation gas and had a landing weight of about 50 tons.

The flight schedule, to which we mostly adhered, called for stops at Paramaribo; Recife (Brazil); Abidjan (Ivory Coast); Fernando Po (later renamed Bioko); Luanda (Angola); Cape Town (South Africa); Mauritius; Singapore; Darwin, Perth, Adelaide and Canberra in Australia; Port Moresby (Papua New Guinea); the Kwajalein atoll; Honolulu; Los Angeles; and Denver, We were scheduled to return to Washington at 3 p.m. on March 2, 1968.

We did not land in Luanda due to the ongoing independence war against the colonial power, Portugal. That put us ahead of schedule and allowed for a bonus flight to and from South Africa. The visit to Johannesburg was not planned but happened because soon after we headed from Capetown to Mauritius, a crew member noticed a fuel leak on one wing. I recall how the fuel from that tank was draining faster than expected. We turned and headed back, jettisoning nearly all fuel. A spectacular sight: the streaming fuel illuminated by aircraft lights. Ambulances and fire engines waited for us on the runways. Turns out this was a false alarm; it was actually a fairly harmless oil leak. But there was insufficient high octane left in Cape Town so we obtained it in Johannesburg.

There was concern about fuel quality at our next stop in Mauritius, so landing on Reunion was considered. However in the end we flew to Mauritius, which I had visited on RV *Chain* four years before.

Nigeria was bypassed as planned, due to their ongoing civil war in which Biafra attempted to secede. While in Santa Isabel on Fernando Po, we saw posters that read: "To Keep Nigeria One, Is a Job that Must Be Done."

The Recife and Abidjan flights were a departure from typical high-level oceanic geomagnetic deviation flights. Instead, we flew over land and coastal areas both in Brazil and Africa to see if magnetic anomaly patterns due to ancient geology matched up when the continents were reassembled in jigsaw-puzzle fashion. On those flights, I worked as a geoscientist assisting Canadian researcher Dr. Dave Strangway, who had contacted Project Magnet in the hopes of getting those lower-level flights. This is the only case in my career where this application was tried. It didn't help much if any due to not enough data and lack of overflight permission by at least one African nation. Strangway and I published the results in 1971.

CHAPTER 4

Western North Atlantic: A History of Ancient Flips of Geomagnetic Pole

At GOFAR, just a year after finding all those magnetic data in the NAVOCEANO Magnetic Division, I began collaborating as scientist coach and coauthor in interpreting the detailed magnetic survey by the ship *Keathley*. That survey covered the deep Atlantic from our eastern US margin east to somewhat east of Bermuda. The survey lines were mostly east-west with a few north to south, and from just north of the Bahamas to about Bermuda. This survey had discovered and mapped a much older set of magnetic lineations than then had been reported nearer the Mid-Atlantic Ridge.

The younger crust was not mapped by the *Keathley*, but other ship tracks showed the younger lineations. On the African side, both younger and older lineations appeared, recorded on various scattered tracks of ships that had towed a magnetometer.

Like a two-sided tape recording, the same set of linear anomalies exists on both sides of the Mid-Atlantic Ridge axis, the present plate boundary. The closer to the boundary, the younger.

My NAVO coauthors and I named this wide band of linear anomalies after the survey ship, *Keathley* (which I never saw or served on).

This and the mirror-image band never before mapped in such detail recorded how fast our plate moved away from the African plate—i.e., the speed of the tape recorder—and the history of magnetic polarity flips during the early history of the central North Atlantic.

Back then for lack of better data, I overestimated the age; today this series of reversals and oceanic crust is known to date from 155 million to 120 million years ago, a fair chunk of the age of dinosaurs.

I presented these exciting discoveries at a conference in Madrid, Spain in 1970.

By 1969 at GOFAR, I had proposed mapping those same ancient magnetic lineations from Bermuda up to the edge of oceanic crust southwest of the Grand Banks. This we would be doing that very year, flying on the Project Magnet Super Constellation or possibly their other aircraft, the P-3 Orion.

Thus (alas, with my family car keys forgotten in my pocket), I flew with the team to Bermuda and from the airbase there flew some of the tracks. With me as chief scientists were Walter Jahn and Allen Lowrie from GOFAR and Len Dennis from the Magnetics Division. The others were the USN officers and crew; only one or two had been on the 1968 flights.

We flew a few more transects of those anomalies and eventually returned from Otis Air Force Base, back to Andrews. Some of the flights were very rough, at only 500 feet, especially crossing the Gulf Stream, which we did on many flights.

I can't recall if we had those new aero-data processed in time for my talk in Madrid. The time was harried because I had a ship cruise scheduled for 1970 on *Lynch*, as you'll read in the next chapter, to examine the crust of the east flank of the

Bermuda Rise, a sub-marine swell later found to have formed by uplift of pre-existing ocean crust some 45 million years ago.

As well as science, the expedition had personal highlights. On one flight when we landed at Otis Air Force Base, Joint Base Cape Cod today. I recall getting together with my brother, then at Harvard, and his future wife Carla, then at MIT, in some eatery for dinner. I think he picked me up and brought me back to Otis.

It also happened that I had learned a Scientists Cliffs cabin owner a stone's throw from our home in Maryland belonged to the US Consul to Bermuda at the time, Charles Manning. I got up the nerve to call the consul, who invited me to an official party at his palatial consulate (later sold to save US taxpayers). The party was on an off night, not a night before another flight. I explained I only had casual civvies and a green flight suit, but he was okay with that, and I had a good time. His son, Michael, and his Bermudan wife live in that Scientists Cliffs cabin today.

Letters Home

Late Oct–early Dec 1969 aero-mag survey of NW central Atlantic between Bermuda and across the New England Seamounts. Same Super Connie as 1968 around the world. This was the first field research project I proposed and laid out in my career. I left Randi and baby Erik at Andrews AFB without the car key.

30 Oct 1969 partly cloudy & 68 F

Postcard from Bermuda

Oops! The keys are on the way, but too late. You no doubt called Ginger to pick up spare key at house & then drive out. Staying at BOQ with W.J. [Walter Jahn], L.D. [Leonard Dennis of Project Magnet] with navigator, two rooms shared bath; Hamilton is too far away (half hour by bus; $5 by cab) & will be arising 2 or 3 a.m., take off before dawn, 15 hour flights every other day, should go quickly. Windy cold front thru Wash day before I left. We overtook cloudy, windy, coolish, typical oceanic blustery weather.

October 31, 1969
From Kindley AFB, Bermuda, Halloween night, 1969

> *Dear Fair-Haired Wifelet:*
>
> *What would you like to hear?*
>
> *There is little new to tease your ear*
>
> *It's stuffy and routine, I fear*
>
> *I would rather be at home, my dear!*

A hard long day (4 a.m.) is behind us, and we just gorged ourselves at the 'O Club.' The bar is quite dead, save for the occasional clang of slot machines legal here. It's still breezy and mild, perhaps 65 F, with the quiet only broken by the roar-whine of an occasional jet transport. The PX, etc. are sparse and frumpy, the bar population only a skeleton of its former self, in the late 1940s and '50s, when B-52s, B-47s etc stopped here to refuel, then once more at Santa Maria, Azores, before alighting in Frankfurt. In fact, Kindley AFB will be turned over to the Navy in July 70.

Our first flight lasted 12 or 13 hours, with a long cross-check dogleg north. We did the whole thing on this one (sextant, mag heading, etc).

Flew at 7500 & finally at 8500 feet, skimming just above the tops of a broken stratocumulus deck. In the afternoon we began the survey legs, dropping down to 1000 feet. It seems like skimming above the water! The scene is bumpy down there and like queasy (yawn). Navigation is our hang-up.

NB: When addressing mail be sure and include PROJECT MAGNET AIRCRAFT 145-925 in the address.

Only one Loran-C station could be received, especially when we descended to 1000 feet & the navigators reverted to Loran-A, which they know better but which is more complicated. Several-mile navigation accuracy is the best that can be expected. One set doesn't work right and nobody on board knows how to fix it.

We're getting along well, the three of us, so far. Walter & Leonard Dennis are affable sorts. The skipper, whom you saw, is a talkative type, for a change, & he tells us about the weather picture, aircraft problems, etc., etc. The chief navigator was also on the March '68 flights, as was one ground mechanic and the plane captain (meaning the CPO who heads the mechanic contingent). Nobody is overjoyed at Bermuda, because the liberty is dull, especially for swingers who can't amuse themselves by writing their wives or girlfriends, say.

We fly again the day after tomorrow (i.e., on Sunday) & the weather should be ok, hopefully not so bumpy. Tomorrow we look at today's data.

The basic operating procedures came back quickly today for me, not too many f—-ups, and Walt learned it too. Leonard is the most experienced active party chief around, knows more I think than Charlie Gunn.

Tell me about the bridge party, and how are you getting along without your man? Not too well, I may hope. Did you know Alexandra [Walter's wife, Slovakian by birth] is only 20? So you're not the youngest any longer. From Walt's description here and there (we don't sit around making cracks about our wives), I would think you may have something in common, at least a European outlook and a taste for Scandinavian furniture. Why don't you commiserate with her some afternoon?

3 November, 1969 9 a.m.

Dear Lover,

Lonesomeness is beginning to set in, but work—hard work—is the best antidote. We're hours stamping and timing the magnetic charts, reading them, converting to gamma values, and graphing against time. I have done little sightseeing, though a few hours will do St. George, and Hamilton, and a few little things to buy. Except empire products, Bermuda is no bargain, but the PX is.

We've been doing our data reduction here in the room, or down the hall on Formica tables in the usually deserted TV room. Last night I drank it up with a boisterous and witty group composed of two of the three navigators & the third and fourth pilots. (We don't need so many pilots but someone put them on board for training.) The youngest, who also came on board yesterday at 4 a.m. with a great hangover, has only 12 hours on the Connie, he confided (hoping the skipper wouldn't find out). Four of us played darts in the bar several hours, drank stingers, & then they started to discuss folk singers, during which I felt stupid, and then they gathered around the piano for a horribly out of tune "Wild Irish Rose"

& the like. Then I left & they followed back, waking the apartments with foul lyrics. I'm afraid I encouraged this carrying on, opening our apt with my key for them. They danced & sang around poor Walt in the sack, then thru the shared bath to arouse Len & the third navigator, a quiet novice, & turning on showers and faucets. They then trooped into the skipper's room (I heard his hard words & the eventual silence).

The trouble with these gay blades is that they basically don't give a damn about the quality of the navigation. The less work, the better. They refuse to use an excellent but a trifle harder to use (more knobs) E.D.O. Loran radio navigation system set on the aircraft. They could have learned to use it before the trip. But, yesterday, when one set gave out, the cross rate being undecipherable, we had to abort (by Leonard's decision) the flight after 4 legs (6 was planned) & return to Bermuda. The Radio officer is sore at the navigators because 'the sets aren't working' reflects badly on him. He got 5 stations on ground with the set that was to have malfunctioned. Tomorrow we fly with a spare. Apparently the skipper also gave them a talking to.

Tomorrow may be rough; the high has wandered east & a necklace of cyclones is moving over the East Coast, seaward. The *Franconia*, once-a-week Cunard liner from N.Y., arrived 7 hours late. You will be hearing the surf, rain & falling nuts at this very time...If you feel lonely, find a project and work hard at it...Hope to hear from you soon. Did you hear how the Escowitz and Treadwell parties went? Cars holding up? What should I be shopping for again? How are shades & curtains coming? (hint, hint)

What else to mention, oh yes, the houses are all pastel colors here, made of a very light dune limestone that can be trimmed with a carpenter's saw. The islands are really just huge cemented windblown dunes from the ice ages when a vast coral platform on the Bermuda volcano & adjacent banks were exposed to the wind. All interglacial (4) soil horizons have been identified in well holes & road cuts, incl. fossil woods & bones. Bermuda is really the tip of a Pike's Peak-sized mountain, 200 miles wide at its base. Volcanic rock is just 250 feet down & it shows on the magnetics!

Well, enough of the natural history. The people here are friendly and dignified, their English is mixed American and empire pidgin traces. Due to stringent laws Bermuda has not been colonized by bastard American culture & architecture, as have some more far away places.

Well it's 10 & I must be up at 3:30 or 4 (yecch). Must remember to eat a piece of bread or something for breakfast before we take off. Otherwise the bumping around on an empty stomach is a queasy matter.

CHAPTER 5

Deep Down, A Lethal Nerve-Gas Cargo

On August 18, 1970, the Liberty ship *Le Baron Russell Briggs,* named after the first Dean of Men at Harvard, was scuttled by the US Army in 16,000 feet of water in the Blake Bahama Basin, 282 miles east of Cape Kennedy, Florida. The vessel's cargo was 418 steel-jacketed concrete containers filled with 12,540 US Army M55 vs rockets—and one additional container. The rockets were filled with sarin, nerve gas. The other container held VX gas, another extremely toxic nerve agent. The *Briggs* was scuttled to dispose of what had become waste.

At the time, I knew nothing of this scuttling. However, this was my first chance to be chief scientist on a geophysical research cruise. At the Naval Oceanographic Office, which goes by NAVOCEANO or NAVO, the chief is known as the Senior NAVOCEANO Rep, or still another acronym— SNR. With my small team of marine geoscientists, we were planning to examine the east flank of the broad Bermuda Rise, a sub-marine swell later found to have formed by uplift of pre-existing ocean crust about 45 million years ago. We were not researching the formation of the Bermuda Rise but the crust itself, formed by the sea-floor spreading at the Mid-Atlantic Ridge during the time between 83 and 120 or 121 million years ago. (This was during the time of dinosaurs, but we were not researching them.)

For those 37 or 38 million years, it appeared that Earth's magnetic field had not reversed—for reasons still mysterious today. We planned to tow our magnetometer and monitor our echo sounder to learn more about this interval. Maybe we would detect evidence of at least a few magnetic reversals of polarity (the magnetic North Pole becoming the magnetic South Pole).

Our research vessel, *Lynch*, fell into the category of Auxiliary General Oceanographic Research Vessel, or AGOR. A number of my research cruises would be on AGOR sister ships—from 209 to 246 feet long and all but one from 1,400 to 1,900 loaded tonnage. They were mere rowboats compared to modern cruise liners, which are up to 1,200 feet long (*Icon of the Seas*), 210 feet in beam and 250,000 tons, with nearly 10,000 passengers and crew.

Lynch pushed off from the Navy Yard, Washington, DC. This was before Hurricane Agnes in 1972, which clogged the Potomac channel with so much sediment that even vessels as small as *Lynch* could no longer reach DC and, in fact, most of the tidal Potomac. There was too much controversy about how and where to dispose of dredge spoils, so Agnes left a lasting legacy.

My wife dropped me off at the Navy Yard dock and drove down Indian Head Highway to follow our first few miles down the Potomac. On our way south to Norfolk, I was handed a radio message informing me that *Lynch* would have to divert to the Briggs scuttle site and try to figure out if the hull had cracked apart (as the *Titanic* did almost 60 years earlier) before settling on the ocean floor, perhaps spilling nerve gas containers. That was the first any of us had heard of the scuttling. The message said that the Naval Research Laboratory, where I would later spend most of my career, would later examine the wreck with more specialized instruments and scientists. We were told to pick up a Lab water chemist PhD in Norfolk. He would test the surface water on board for any sign of nerve gas escape.

Upon departing Norfolk, we headed straight to the scuttling location. However, that was before GPS so there was some uncertainty about the exact spot. Moreover, the *Briggs* might not have gone down vertically. *Lynch* only had a conventional wide-beam, 12 cps echo sounder (12 cps means the pings we emitted were sounds at 12,000 cycles per second) which works on the principle of transmitting sound waves from ship's bottom and then measuring the time taken for the echo to be returned from sea floor. Human ears pick up sounds between 20 and 20,000 cps. The outgoing pings were clearly audible in the ship's laboratory in the main deck. Because sound in the ocean travels at about 1,500 meters per second, the pings have to be spaced from 5-to-10 seconds apart at ocean depths, to allow time for the ping to reach the seafloor and echo back to the ship.

Wide-beam means that our echo sounders emitted conical beams of sound up to about 30 degrees from the vertical. Ideally, echo sounders record only the ocean depth directly below the ship. Advanced echo sounders are narrow-beam and do that, and modern multi-narrow beach equipment creates and records a swath of sound, so modern research ships 'mow the lawn,' mapping a swath of seafloor topography.

We crossed the reported *Briggs* site several times but did not notice anything. Stopping the vessel several times allowed the

Naval Research Lab chemist to collect water samples, which he found contained no evidence of nerve gas. I am not a chemist but was very skeptical about the chance of detecting any, even if the *Briggs* had broken apart.

However, the sea was fairly calm, so I decided to try just drifting across the reported wreck site. Lo and behold, a local dark smudge appeared on the record. However it came in after the bottom return, so maybe it was a buried object like a large boulder. Or the return was a side echo, which would appear as if it was from a buried object. We drifted *Lynch* across the reported location and each time there was that smudge. Only one smudge on each pass. I plotted the locations of the smudges at our *Lynch* positions and they were scattered around one area near the reported scuttle location. It became obvious we had found the *Briggs* and that it had not broken.

The key was to drift slowly, so there were a few echoes from the wreck from each pass. In 16,000 feet of water, it takes around six or seven seconds for an outgoing ping to echo back to the ship. If the vessel travels at full speed, there wouldn't be enough echoes from a wreck. A ship steaming at 15 knots would cover about 1,500 feet between pings, but drifting at 0.5 knot only 75 feet. This was before GPS, so we could not know our precise drifting speed. Liberty ships like *Briggs* were 441 feet long and 57 feet in beam. The pings were not thin fans, and the ship was always pitching, yawing, and rolling a bit. So there is no doubt that a number of pings would 'ensonify' the wreck and reflect back to our *Lynch*. Thus there can be little doubt we got echoes back from the wreck. Our best, strongest, and longest echos would be when *Lynch* paralleled the wreck. The weakest occurred if the wreck lay perpendicular. But we could not be certain of that. Probably it was a mixture of in-betweens.

I wrote up my conclusions. Briefly docked back in Norfolk, I was met by a young Naval officer. I handed him the report and data.

As a postscript to our *Briggs* data, I learned that this Navy officer had misinterpreted my chart with the locations of smudge marks. He had appended his report, claiming we on *Lynch* had found that *Briggs* had broken into pieces.

We had used a wide-beam echo sounder as a primitive side-scan sonar to locate a wreck in 16,000 feet of water. It may well have been the deepest shipwreck ever located that way— and in fact, it was 4,000 feet deeper than the *Titanic*. The latter had, as we all know, broken into two large parts, the longer middle-to-bow section longer than the *Briggs*. I wondered if *Titanic* could have been detected that way—or was detected, but not noticed.

Robert Ballard rediscovered the *Titanic* wreck in September 1985 on board the Woods Hole vessel *Knorr*. His team visually identified *Titanic* debris on videos on the deep-towed Argo robotic system—of course no echo sounding could have done that. Out of curiosity I contacted (emailed) both Woods Hole Oceanographic Institution and Ballard, now a professor at the University of Rhode Island. The Woods Hole data librarian kindly provided links to digital data and archives. I could not find evidence of the *Knorr's* echo sounding analog records. Neither of my two emails to Ballard bounced nor were answered. So the jury is still out as to the question of echo sounder evidence for the *Titanic* wreck. Of course an enterprising reader of this book—someone with the right connections—might get a vessel with a wide-beam echo sounder to slowly drift the ship past the exceedingly well documented *Titanic* wreck—arguably the most famous in the world. Of course, maybe the sea floor is too rough at the site with too many side echoes from ravines or large rocks.

As for disposing of hazardous waste on the ocean floors, it's probably highly illegal now and not done any more—at least not legally. The USSR disposed of radioactive waste in the ocean into the 1980s, as will be touched upon later in this book.

It should be noted, though, that losing a nuclear submarine is a tragedy, not intentional disposal of toxic waste. All that is afterthought.

After the data handoff in Norfolk, we headed back to sea and collected data over the Bermuda Rise. Our findings were published in *Nature*.

CHAPTER 6

In the Indian Ocean, Running the Show

With colleagues William F. Ruddiman, Linda Glover, and Fred Bowles, I applied for and obtained use of the *Bartlett* for several months of research in the far-western Pacific East Indies island arc, (Indonesia), and eastern Indian Ocean south of Indonesia. *Bartlett* was an Auxiliary General Oceanographic Research (AGOR) class research ship like the *Lynch* I had used back in 1970. The period late 1971 to early 1972 unfortunately required some of us, me included, to be at sea over the Christmas and New Year holidays. My wife Randi was not pleased to be left alone with baby Jason and brother Erik, a first-grader.

Fred Bowles was a sedimentologist who took seafloor cores, sampling them later for sediment size, type, and especially clay mineralogy. Bill and Linda were micropaleontologists, examining sediment cores and analyzing under microscopes planktonic and benthic (bottom-dwelling) kinds of foraminifera, diatoms (tiny floating plants), and nanoplankton (tinier floating plants). These tiny fossils had keys to ancient ocean currents, temperatures, and global glacier ice volumes. Due to the very slow 'snow' of sediment—including microfossils—on most deep ocean floors far from land, a 10-foot-long sediment core could have sediment ages from zero at the core top to hundreds of thousands of years at the bottom. At the bottom, a crown-shaped core catcher was attached to the pipe-like core barrel to keep the sediment from sliding back out when the core was pulled up to the ship.

I took over as *Bartlett* chief scientist in Djakarta, Indonesia. We had one port stop in Darwin, Australia, and our second leg ended in Guam. Of those who disembarked there, Ray Bracey flew home to the States, but I had committed to presenting a paper (i.e., a short talk illustrated by 35-mm slides) at a geological conference at Victoria University, in Wellington, New Zealand. I had to fly to Honolulu first to get from Guam to New Zealand.

Chief Scientist Vogt digs rock samples out of pipe dredge.
East Indies Bartlett December 1971.

For my part, I was interested in mapping linear magnetic anomalies in Wharton Basin, a deep part of the Indian Ocean south of the even deeper Java Trench, where the Indian plate is sinking below the East Indies along the Java trench. At the time, the Wharton Basin was considered possibly underlain by the world's oldest surviving oceanic crust. The Indonesian volcanic island arc, produced by the subduction of the Indian plate, includes some of the world's most infamous volcanoes, notably Krakatoa and Tambora. In December 2004, a 9.1 magnitude earthquake under and along the Java Trench generated a tsunami that killed more than 200,000 people. The popular and pre-plate tectonic media term 'Ring of Fire' around parts of the Pacific includes the Indonesian island arc.

For my turn as *Bartlett* chief scientist, I invited two accomplished geoscientists along. One of them was Dewey (Ray) Bracey of the Naval Oceanographic Office magnetics division. Ray had some new interpretations of the Caroline Ridge, a relatively young but not well-known seafloor swell north of New Guinea. I gave Ray a chance to test his interpretation with a *Bartlett* magnetic track partly of his choosing. My other guest scientist was Professor John Connolly, an Australian geologist then at the University of South Carolina. John had some hypotheses about tectonics and volcanism inside the complex Indonesian archipelago. We could possibly test these ideas by dredging rocks in a deep basin. Today I cannot recall his ideas and the research issues involved.

Bartlett and her officers and crew were docked in San Diego at the time Global Ocean Floor Analysis and Research (GOFAR) took over, so the first team flew out to install instruments and embark for the first leg—crossing the wide Pacific. A young law student was on the team. He had worked a summer job at GOFAR in Randle Cliffs and wanted to see a bit of the world and learn some science before pursuing a less adventurous career as a lawyer. As required for all who shipped out on Naval Oceanographic Office (NAVO) vessels, he had to pass a physical exam. He was in great shape but had one problem—diabetes. NAVO and all Navy prohibited diabetics from going to sea. What if they got sick and couldn't keep down the sweet stuff to go with the insulin? He pleaded with GOFAR head Lt. Eric Schneider. The young man declared he had been on small boats and never got seasick. After some discussions among NAVO and other Navy staff, it was decided that the man's private physician could make the decision to override that of the Navy doctor. And so our young diabetic boarded *Bartlett* with the others.

The Pacific was named by Balboa because it looked so calm, but that was in Panama. Off Southern California, the ocean is also calm, particularly in summer. Now it was fall and while there was no major storm, *Bartlett* headed and pitched west into significant seas once out of the harbor. After a few days out, we at GOFAR got radio messages that the law student was seasick and couldn't keep anything down. There was no doctor aboard. It did not look good, so *Bartlett* was directed to turn back to San Diego. The young diabetic's condition worsened en route, so a helicopter flew out to evacuate him back to the US for emergency hospital care. He died onboard the flight.

We back at GOFAR were in shock. We had never had fatalities or even serious injuries at sea before. Schneider was especially despondent, as he had helped get the law student cleared to participate. Now he had to fly to Hawaii and talk to the young man's mom. The death of this young bright law student placed a persistent dark cloud over the relationship between us senior scientists with our boss,

Eric Schneider, who felt we had not stood by him in this emotional ordeal.

I imagine that this incident resulted in more strict rules and no exceptions when it came to taking diabetics on ocean cruises. US survey ships like *Bartlett*, as well as university research vessels, rarely took medical doctors to sea—probably mainly because medical professionals run up expedition costs a lot. Could this fatality have been prevented if *Bartlett* had had a medical doctor aboard? I don't know. Maybe the ship would have turned back sooner. When someone got seriously ill or injured at sea, it was generally the first mate who was instructed by radio on what to do. First mates may have some first-responder training, but taking out an appendix several thousand miles away from civilization? What if my wisdom tooth impaction had happened on an ocean cruise? Some people advise ocean-going folks to have healthy appendixes and wisdom teeth removed as a precaution. I still have my appendix and will likely die with it, not because of it.

On most of my ocean expeditions, there were no doctors on board. An exception was 1986–'87 in the South Atlantic. An American doctor came along for free, just for the adventure and experiences. Vessel port stops in Recife, Montevideo, and Capetown were likely inducements. He was an avocational gemologist who explained the Brazilian and South African gems he and some of us had purchased. He stood some of the science watches as well, upon coaching.

I recall the Russian ship *Keldysh* (with its Mir submersibles, see later in this book) had a doctor on board. That was 1998, and he probably earned far less than a US counterpart. Those of us who planned to dive had to pass a physical on board. Do you feel okay? Do you suffer from claustrophobia? Only one brilliant Russian scientist, Gabriel Ginsberg, was denied permission to dive. He was frail and suffering from late-stage cancer. In fact he died not too long after our cruise, and I wrote his science obituary in memoriam.

I took over *Bartlett*—from Ruddiman and Glover—in Djakarta, Indonesia. I bought a finely carved (ebony?) mask as a souvenir: a Hindu goddess (Ganesh) with an elephant-like trunk. Years later, I gave it to my younger son, Jason, and it hung on the wall of our family house in Santa Barbara. My other souvenir from my stint on *Bartlett* was a set of two giant water buffalo horns. The original sharp tips had been sawn off or rounded off. While domesticated, water buffalo were widely used as draft animals in Indonesia, I acquired these horns in Australia, where *Bartlett* made a port stop in Darwin (where I had visited five years earlier on the Project Magnet flight around the world). Water buffalo were imported to northern Australia by English colonists and now ran wild.

At least in the NAVO-Navy world, it was expected that someone responsible stood watch on the deck of a docked vessel. It's called showing the flag. Visitors or port authorities might be allowed on board, and various supplies, and in our case scientific gear, need protection. One day when it was my turn to stand watch in Darwin, I was leaning on a railing overlooking the dock when I saw a man peeing into the harbor near the ship. Pretty brazen and disrespectful I thought. After he had finished, I shouted down to him and we began to chat. He was an American expat and was happy to see a US ship. I can't recall if I invited him aboard for a quick tour, but he ended up inviting me to dinner with his family in their Darwin home.

At the end of the dinner they presented me with the two buffalo horns. Maybe his wife was pleased to get them out of the house. Today my wife would appreciate getting them out of our house. We have two sons, so one buffalo horn for each? They would hold gallons of booze or many small gifts.

In my memory, no great science resulted from that cruise. The oceanic crust under the Wharton Basin turned out not to be very ancient, but to get younger closer to the Java Trench. A branch of the Mid-Oceanic Ridge had evidently been swallowed. The rock dredge suggested by John Connolly had returned few rocks, all of little importance. I recall that Connolly scornfully dismissed the typical US teabag tea served in the galley. He asserted that tea bags were filled with inferior leftovers swept off tables. Ray Bracey did get some useful survey data over the Caroline Ridge—and published it in 1975. It would have been ethical to add yours truly as co-author.

Letters Home

1971–'72 Bartlett cruise two legs eastern Indian and western Pacific. Get on Djakarta, Indonesia; Port Stop, Darwin, Australia; off in Aduana, Guam, fly via Honolulu to Wellington, New Zealand to give paper on cruise findings at conference Victoria University, Wellington, NZ.

4 December 1971
Postcard enroute to join ship
Sydney, Australia, 9 a.m. (Santa Barbara Day)

Dear Randi,

Hey, here's another word from me. Now your day is my night and my summer is your winter. Between us now is only the mysterious Earth, its insides unseen by man...

A chill wind and low clouds came in suddenly to evict the heat of an early summer day...lavender jacarandas and other lovely flora also planted in California and such places, and shrilling cicadas, and chirping birds say it is early summer. Sydney has LA's sprawl, SF's setting and appeal—lots of blue embayments—and Washington's population, but it seems cleaner, friendlier, less sinister than any US downtowns. Wanna emigrate? Sydney is also a great melting pot. In my dazed wanderings today I must have heard half a dozen languages.

Randi please remember to pay Messenger $500 [holds mortgage on our house] and A. Lowrie $25.

AIR MAIL
PAR AVION

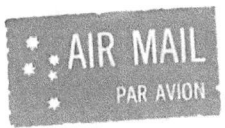

6 December 1971
Postcard from Djakarta, Indonesia

Hello Lover,

This will be last for a while, we sail 1500, unless skipper gets bad diagnosis—apparently had a mild heart attack two days ago...The appearance of Ruddiman et al [colleagues who had the previous legs waiting for their flights] at airport a most warming surprise; camaraderie grows best in remote places...Djakarta out South Americas South America...Too, too bad no time to wander around and absorb...

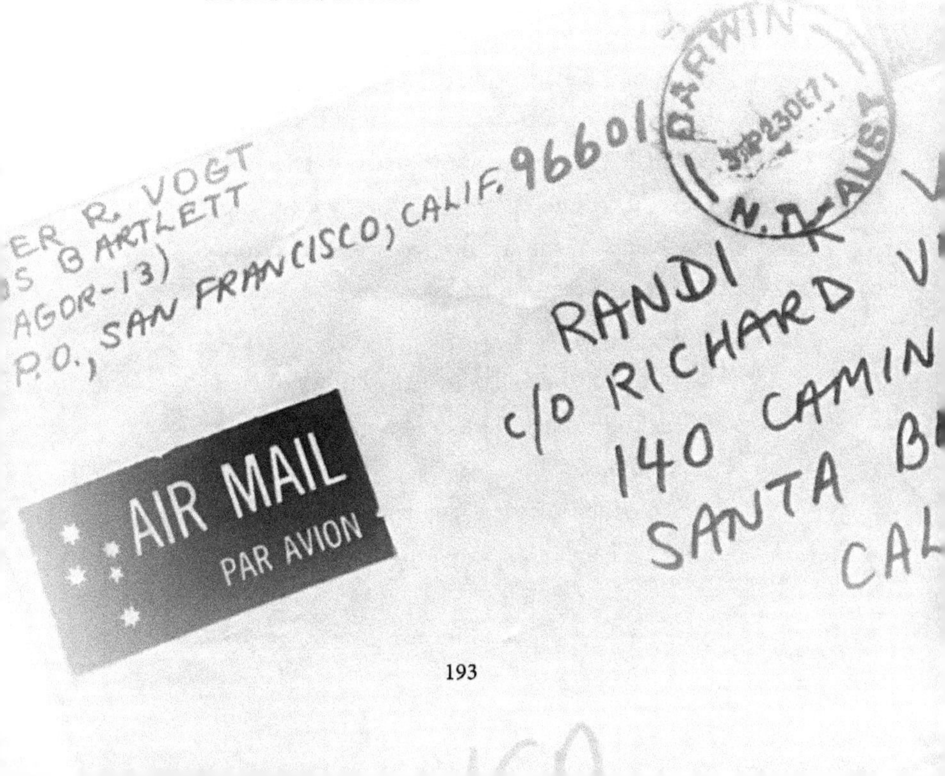
ER R. VOGT
S BARTLETT
AGOR-13)
P.O., SAN FRANCISCO, CALIF. 96601
DARWIN
N.T. AUST
RANDI
c/o RICHARD V
140 CAMIN
SANTA B
CAL
AIR MAIL
PAR AVION

16 December, 1971
Letter later posted from Darwin, Australia
About 11° S, 105° E, south of Java and heading for Darwin

Dear Randi,

The sea is blue and the tropical sun blazes down, so how about a few lines from me? The drone of day to day routine makes it hard to imagine that somewhere beyond that razor-sharp edge between the dark blue disc and the light blue bowl of sky there are other people, loud cities, holiday shoppers, snow, and a lovely fruiting wife, now dreaming in California, on Earth's nightside, perhaps I may dream it's of me she is dreaming. A harrowing (but not too much so) trip behind her, to relax with good people [my parents]. I don't worry too much but feel that everything is going well, and that your sadness of separation is always silvered with our reunion, soon and certain and at least as sweet as memories of past reunions.

Right now I count the days (6) until your words and thoughts are in my hand, Christmas companions. The cruise is going well so far, and morale is rather good among our long timers (Pennylegion and Scarborough) [already on ship-probably since San Diego, when my team got on in Djakarta], and the military sea transport service officers and crew, most of whom are cooperative, even interested. The captain is a spare, calm man, quite the opposite Old Man Hobbs of the *Lynch*. *Bartlett* is basically the same ship, but nicer in small ways—a large flying bridge, with several deck chairs for sunny days. There is a scientists' lounge, although used for movies. Jim Pew has shown me how to run the 16mm projector and I have shown several movies, to crowds of 5 or 6. We each stand 4 hours watch, spend a few hours plotting up data (Jay and I do this mostly) and that's basically it. Eat, sleep, shit, shower, but no, no more shaving since I'm letting it all grow out, the way I was in the good old days.

The galley is a trace pleasanter than *Lynch*, the food passable but unexciting, but our short-goateed, Black mess 'boy' Shorty is rather a riot. Ask Ruddiman.

I have eaten lightly and believe to be losing some weight. Food is served 7:30–8:30 a.m., 11:30–12:30 p.m. and much to my dislike 4:30–5:30 p.m.

Bracey (Dewey "Ray" Bracey, whom I invited along from the Magnetic Division to give him a chance to get into the science) is doing it all too, with his same cynicism about everything. I hope instruments hold up after Darwin, where Pew gets off, for Ray's sake. Bracey is suffering from constipation; lack of exercise, probably the cause, as is shown by the fact that his automatic rewind watch ran down. He also may be suffering from beer-withdrawal, since he told me he averages 6-8 beers per evening. Maybe he can get good and tanked up in Darwin...

Getting to know Jim Pew better, many long bull sessions I really enjoyed. Too bad he's getting off but who wouldn't after four months.

(Jim also introduced me to the ship's machine room which I had supposed off-limits for hobby work. I learned to make and still have brass belt buckles and pendants from brass stock.)

Last night we even had a little party, Pennylegion, young Scarborough and Pew (the 'in' group) and they invited me. Crowded into Pennylegion's stateroom we listened to tape, guzzled gin (booze on ship illegal but not rigorously controlled if people are discreet), nuts, some gouda cheese from the galley and a finger-printed red clay that turned out to be fudge that Wally's mom had sent him. Pew told Arctic stories and people wishful-thinked about getting our own little GOFAR schooner cheap and fired on spirit de corps and GOFAR competence, with wives as cooks, and all that. I felt very warm and social; back in the lab months might pass without a good morning or other trivial remark. In a small way that reality silvers the droning, grey, automated, and isolated existence out here.

Meanwhile the sea is lightly ruffled with rolling swells, like the flexing muscles of a smooth-scaled monster, pitching our ship about. The first half week out of Djakarta was choppy, cloudy, monsoonly, and I felt rotten. I woke up often very dizzy, like at home only worse. Considering it was not really rough, adjustment seemed to take forever, and cost a few stomach-fuls. Now I'm ok, but this is soft, not North Atlantic. People leave cups, pencils, lab books and tools lying around on the work tables.

Think hard to remember what *Lynch*'s lab looked like: This is the same. [Randi toured *Lynch* tied up Navy Yard before our 1970 cruise on that ship.] Getting burned up, so much for now.

20 December, 1971

Dear Randi,

A two-day attack of bad dysentery behind me; the ship must still be crawling with the stuff. Only three days out of Darwin now & up on the Australian continental shelf. Water and air on the rise-up to 86 F, pretty hot for the open ocean, & air conditioning can't hack the sun on the steel decks anymore.

Hope you will tape over the personal touches in this letter & let my parents read it too (no point repeating everything). I realize you won't be reading this until after Christmas, but I'll be thinking of you all at about noon on the 25th—while you celebrate Christmas Eve. Is that right?

TER R. VOGT
NS BARTLETT
AGOR-13)
2.O., SAN FRANCISCO, CALIF. 96601

DARWIN
23 0E7
N.T. AU

RANDI
c/o RICHARD V
140 CA

23 December, 1971

Dear Randi,

A clear hot humid day. At anchor off Darwin. No pilot yet and no health official. Egloff forgot his shot card; if we don't get off, someone will get shot! Thinking of you often, sometimes too intensely at night. We do have a few touches of Xmas on this ship: in the galley some colored flashing lights, glass balls, and a 1½ ft plastic tree with some sorry apples and oranges under it; an inflated Santa next to the usually useless TV— and a 7 ft plastic fir bought in Djakarta believe it or not, will be strapped to the top of the mast. It's a tradition. Lots of love from me to all.

8 January, 1972
At sea some 100 miles north of western New Guinea.
10:20 Australian time

Dear Randi,

Happy New Year, Lover woman!

At last the seas, the body and mind-debilitating seas, have stopped tossing this $#@*& metal cork around long enough for motivation to rise above necessity. It just wasn't very smooth swells coming from this way and that. A nearby depression turned into a storm in 24 hours and a typhoon in another 24 hrs, fortunately it's off near the Philippines now.

Bracey is finally off the antibiotics he has been taking for (?) pneumonia the last 10 days; he has wasted away to a cursing, pale scarecrow. Today he plotted some data, ate some shit that passes for food in the galley, and seemed a bit better. Only 1 cigarette the last 2 days, I can't get over it. Even in the depth of his 'pneumonia,' in between hacking, ghastly coughs, our cabin would flicker in the middle of the night with a cigarette's light. He might get chest X-ray Guam and might go home then.

[personal omitted]

[At two and a half months,} this cruise is 2 ½ times too long to be reasonable for me (and you) & you can rest assured that I will never leave this long again.

[personal omitted]

I dreamt I talked to the new baby [Randi was pregnant] & taught him to pronounce new words. He learned quickly and was a perfect prodigy. Then I awoke and worried about rubella & defects & all that, the way one worries far away and half asleep. I hope you have taken extra special care of yourself to keep in the best possible health.

By now you are wading in Wisconsin's snow and Erik is eating it up. Also hooray the trip is half over. Jay [my brother-in-law with whom she and son Erik are staying] would like to go out on a cruise? A month at sea and a reunion like ours is a good spice for a good marriage, love, abiding affection, or whatever different cultures call it.

As far as mistress science is concerned, out here she frequently seems a foul old hag, but I will never abandon her; and your lovely and loving poem on the subject means we can let her 'live' with us, rather as a dear, interesting but crotchety old aunt, a family member, along with the spirit of the wild asparagus, Gamle Norge (old Norway).

[personal omitted]

....Another resolution I made to myself was to take you out more often, even dancing. No, I wouldn't dance with 'ladies of the street' in the situation described in your last letter. But I might dance with you, with a little prodding.

Part 2, 10 PM Guam Naval Station

Dear Randi,

Sorry for this garish pen; that's all there seems to be on this ship. We used these to mark records since they don't photograph in black and white. *Bartlett* lies tied up to a weed & concrete loading area that seems deserted. There are some sub-tenders and a fuel barge, but the place seems like out of *On the Beach*.

Of course we got here after 4:30 p.m., hence everyone is at home. There are no peddlers or artisans. The Guamians have given up what they may have had to live off the American fat. Agana, the capital of Guam, is 15 miles from here; I hitchhiked there this evening, had 4 beers and 40 thoughts & came back to the ship to find MAIL. Agana is Oxon Hill in a tropical setting, perhaps a bit poorer. *Bartlett* is motionless, her rain-wet decks glistening in her lights. Only a handful of people on board. I sit in the galley. The TV, seeming now benevolent and evoking a spirit of home, blares some nonsense from a local station. Your letters and the office garbage is spread out on the formica galley table. I can curl up with your lines and stay on the ship—the hell with ports! No, tomorrow I have to post things.

Wait, Bracey just staggers in. 8 beers in the O-Club. "Fxxx this goddamn ship. Nothing in this Fxxx icebox but the usual rat shit." He is arguing with the assistant cook, so I retreat to our 'state room.' With the ship tied up and motionless (altho it still seems to move) even the dull gray of our surroundings has a silver lining. And then there are your letters (Don't apologize for writing too much, apologize for writing too little!) Your first letters I must have read 5 times all the way through. Nothing you write is trivial.

The coordinator [AGOR coordinator Van Atta] just came in. Wally Scarborough of our group is in the local can for drunken disorderly conduct. Will have to try to get him out tomorrow. Midnight. Shit. I don't feel romantic now. Will write more later. Please please write once more so I'll have mail from you to read on the New Zealand trip. Love, Peter

Undated
Postcard from Guam showing
'Point Two Lovers Leap,' Tumon, Guam

Dear Mrs. Bulgebelly,

I love you! To prove my love, after seeing Wally out of the can, trudged 6 hrs through coral-sand and beer-canned highway shoulders, reaching the 400 ft cliff top just before an anticlimactic rain shower made me dash into a little wooden box labeled 'men.'

Dusk came. Great 9' surf stormed the rocks far below, as the Japs whose pillboxes still picturesquely pock the beaches did, from the north. I hitched 'home.'

Expect me, tired but happy, and anticlimactically in the morning

Saturday the 5th
Leaving Guam

After eating with the whole gang and the Captain at the Red Carpet overlooking 'beautiful downtown Agana' (as Bracey sneeringly puts it) we rode far enough with Van Atta to realize he couldn't find the airport, so we went by cab. Good riddance, Van Atta! Yes, Allan Lowrie, the fellow is a terror! He dashes off in his rented VW, and we go to the airport. Of course, the 747 is delayed (keeping my 100% delay malfunctioning rate on 747s). Bracey has a headache. Bracey is a headache. I have a bellyache. The Red Carpet, Egloff's idea, was a real loser on food and service, offering only high prices and plush trappings. It served only to reinforce my penchant for El Cheapo cafeterias, a reinforcement you could do without, I am sure.

Pan Am from Saigon and Manila finally roars in, and we take off at midnight instead of 730. The plane, you guessed it, is packed. Many militaries and dependents back from 'Nam.' According to my tickets, made out by Pan Am Agana, I arrive in Honolulu Saturday morning, having left Agana (Guam) Saturday night. Then I wait in Honolulu for Saturday night to race around the world and catch up with me.

Change of Plans

Instead of flying home from Honolulu I first flew to New Zealand to attend a relevant science conference and reported on some of our brand new results from Bartlett. The timing was fortuitous.

I boarded American flight 71 leaving Honolulu at 9 a.m. for Auckland. I am to get to Auckland Monday morning. Yes, in all this hide and seek with the International Date Line, Sunday the 6th was the day that never was.

Standing in line for US Customs in Honolulu, Bracey and I grunted our farewells and went our ways. That bastard is home with his wife already! It made me mad to have to go through Customs since I was transiting. I guess almost no one transits from one foreign country to another by way of Honolulu. The D.A. man raised his eyebrows over the water buffalo horns peering out of a knot of graying underwear, but I got them through. [They were a gift from a US expat I met in Darwin.]

I took the suitcase to American Airlines, the man checked it upon inspecting my ticket, and I haven't seen it since. After trying for a few hours to rest in a nearby motel for $10, I came back, checked in, and found the flight doesn't exist. By then my red suitcase was already loaded and flying to Sydney, Australia. I was referred to BOAC, who got me the last vacant seat on Air New Zealand's 551, leaving midnight for Auckland. The man told me the flight Pan Am had booked me on from Auckland to Wellington also didn't exist. There was only one flight a week! Later he consoled me with the fact that surface travel only took 3 hrs. Fortunately, he was full of shit, as I learned from the man sitting next to me when we finally boarded (also an hour late). Surface travel took 12 hours, but a local NZ airline flies often, so I could get off in Auckland with only a two hour wait there.

The flight from Hawaii to Auckland took 8 hours 40 minutes, (about twice the Dulles-LA route) and it was my second night without sleep. My seat was next to the 'kitchen,' and next to the queueing line for the inadequate number of toilets. There were 3 stewards and 2 steward-esses—*pshaw!* People bumped into me all night, and bells rung by passengers summoned the stewards. What assininity! They can use lights for that. But somehow the time goes by, and then the early rising summer sun blazes through the portholes, and the steward dashes into action with warmed up ham, eggs and potatoes, Burp! While all this is happening, all the passengers rise in a chorus of bowel and bladder movements.

Somehow the morning sun and a caffeine infusion set my body on 'daytime' again, even without sleep. It's amazing. All these words of mine will call up replays from your own experience. I would have hated to take this trip with an infant or toddler type! Gadzooks! But in all the harassment I have no duties but to exist.

That's where it differs from, say the ship. Also I plunge from a totally scheduled, somehow secure routine into a totally unscheduled (I mean unrepetitive) world, a kaleidoscope of lights, motion, torrents of strangers. They're both travel yet these two are so diametrically different, it produces a cultural shock to plunge from one to another.

But enough. Even after a 4 hrs nap, I've had it. Tomorrow is registration. It will be small (50). Already saw people I know—Brackett Hersey (the Mogul of Maury Center), Alex Malahoff & Dale Krause, the latter an affable bearded Neanderthal of a man. (Hersey had been chief scientist on part of the 1964 *Chain* cruise, and Malahoff, Office of Naval Research, would get me as guest scientist on a Soviet ship in 1975.)

Then I've got in front of me a stack of papers to read, and my own paper to write. I ate dessert (writing to my lovely wife) first. B. Grosvenor, Barbara, our GOFAR illustrator, sent me slides [35 mm slides for my paper at this conference] & a note on how envious the other women were of how beautiful you looked at the party. See, I keep informed by many channels, and Calvert County, on the other side of the world, is really not that far.

(personal stuff omitted)

You really flattered me in your last letter. We should always remain responsive to our [surroundings]. So many people I know are simply dead, anesthetized, insensitive to 99% of the experience of being alive. They think there are only a few highpoints to anticipate, and existence in between is a nuisance-time to be killed, as conveniently as possible. So little time before we shuffle off our mortal coils, and they talk of killing it! But life is like a mountain hike, the dew and velvet grass and dandelions of the lower pastures should be savored with the joy of posing on the highest, iciest pinnacle..

Randi, I'm happy you bask in my glory, but after all, it's no more glory. After all it was merely publicity about some prior event, not a new achievement. (Must have been some article about me in a local paper.)

Anyway, I'm not ashamed to appear in the small town rag—presumably the *Calvert Independent* or the *Calvert Recorder*. At least it will be read by a number of people who know you and me. A big splash in *Time* can't claim that! I love to lay a laurel or two at your feet now and then, as befits a queen! Only we have to keep the laurels out of our bed. They are even less pleasant than cracker crumbs. With that little tidbit, love to Erik (I do miss the tyke) and see you soon. Good night.

7 February 1972
Writing from Wellington, New Zealand.
9:30 a.m., Weir House, Victoria University

Dear Randi,

Here's another package of words for you. The scene has shifted far away.

The yellow-painted cement blocks squaring off my new home are warm and friendly. Somewhere a toilet gurgles. The cable trolley is briefly there, clattering like a mature train, but subdued. Then the deafening silence returns. Deadening? Serene! Beautiful! Loneliness and silence can be a warm secure blanket. I had forgotten silence.

The room is ascetic by Howard Johnson's standards and Bracey would be pissing and moaning about it. A bed (narrow), a narrow window looking out on lawn and shrubs, and the trolley track. A desk, a drawer chest, a wall closet. No water or toilet, but even so it compares favorably with the Captain's quarters on *Bartlett*. Weir House is a monastic neoclassical affair, complete with student dormitories (this room will be occupied by some student, probably as long-haired as any American, come 20th of this month, when vacations end), conference rooms, and a cafeteria. It's all familiar to me, even though I've never been here before...

February 14, 1972
Postcard showing Victoria University reporting
a tourist outing. Last mail home from
Bartlett cum Wellington

Dear Randi,

The *Moaroana* begins to roll as she clears the sharp rocks, black with mussels and seaweed, the graveyard of her predecessor the *Wahine*, dashed and sunk only a few years ago with many deaths in a sudden hurricane force autumn storm, in early April.

The wind is chilly and the heavy swells are green. This is a different world: the Southern Ocean, and Wellington doesn't let you long forget it; it sits on the edge of the Roaring Forties. *Moaroana* wallows across Cook Strait and into a long drowned river valley, cupped by grass and brush-covered mountains. Some valleys are heavy with smoke, not volcanoes but burning brush and grass to improve the sheep pasturage. Picton is wind-sheltered, even a few California palms, and soon the ferry returns

I talk agnosticism to a game old nun, and every-thingism to a young couple emigrated from Northern Ireland.

Today, Sunday, at last a sunny still day I lazed in the Botanical Garden where the crickets buzzed a Maori word—*waikakariki, waikakariki.* I wandered around and heard a concert below but couldn't see them. My invisible clapping must have cheered them...talked to a Swedish steel salesman and soon this day was over, too.

Paper (talk presented at conference) went tolerably; again one of the major dividends is day to day informal association with colleagues I would otherwise never know.

Little Postcard: will I beat my ass back to my loved ones?

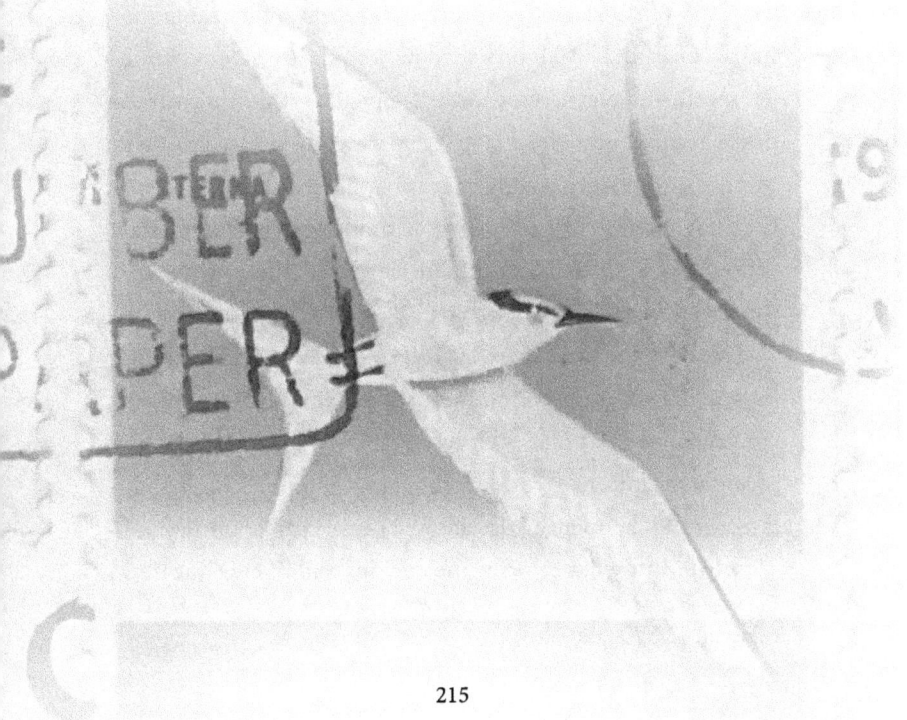

CHAPTER 7

Voyage to Darkness: Rendezvousing with an Eclipse of the Sun

One day in early 1973, I was sitting in a waiting room idly leafing through a magazine—maybe the *New Yorker*. My eyes were attracted to an ad promoting a tourist ocean liner cruise to intercept a solar eclipse in the Equatorial Atlantic west of the Mauritanian coast. How cool!

The total eclipse would occur on June 30, 1973 and sweep from the Guyana-Brazil border across the Atlantic, crossing the Cape Verde Islands into Africa, over Kenya, and end in the Indian Ocean. The Moon's shadow would race at supersonic jet speed—in fact a Concorde (remember this short-lived passenger jet?) was to carry passengers along at the shadow's speed to experience a longer eclipse). As astronomer David Levy has noted, a solar eclipse is a 'natural event with unnatural consequences' and to the fortunate viewer 'has the power to rip through to the core of your being.'

The P&O Cruises liner *Canberra*—with space for 2,600 passengers—was chartered by Eclipse Cruises. This eclipse would be long; in places along its 100-mile-wide swath up to eight minutes of total darkness. No solar eclipse that long would happen again anywhere on Earth until June 25, 2150. But what to offer the curious and likely educated passengers for the rest of a nine-day ocean cruise? It would have to be more than booze and gourmet meals. Eclipse Cruises, Inc. would feature Science at Sea and Culture at Sea programs: lectures on seabirds, astronomy, meteorology, marine biology, space art, NASA's Apollo Program, and more. But wait—ocean floor geology?

Having begun to research the Atlantic and its origin by continental drift and then new plate tectonics, I could be one of their lecturers. Back home, I typed a letter to Eclipse Cruises, offering myself as lecturer in return for free passage for me and my wife. This was a very long shot, and I figured I would not likely get a reply. However, I did eventually get a reply—but it was of the sort I expected: "Thanks for your interest Dr. Vogt, we have had many such offers, but we already have a geoscience lecturer."

Cunard *Adventurer* (also known as *Sunward II Adventurer*)
was a cruise ship built for the Cunard line.

I thought that was the end—nice try, Vogt. But I was wrong. The *Canberra's* cabins were booked in no time. There was much more interest than the company had expected. So they chartered another liner, a bit smaller, the 485-foot MV *Cunard Adventurer.* This vessel would start from North America (Puerto Rico) and intercept the eclipse farther west in the Equatorial Atlantic. On the voyage, the *Adventurer* would offer passengers port visits in Martinique, St Lucia, and Trinidad. Was I still interested? Why of course! I had enough leave days saved up at the US Naval Oceanographic Office.

There would be one other geoscience expert on *Adventurer*: UCLA Professor Howell Williams, then elderly and retired. His expertise was volcanoes, especially those forming the Galapagos. He was the only person aboard whom I knew and was no doubt on board to help passengers understand Caribbean Arc volcanoes, like those on St. Lucia and Martinique. By 1973, it was well established that they result from Atlantic crust and upper mantle forced or sinking under the Caribbean plate and partly melting.

Randi and I made provisions for our year-old son Jason and his 6-year-old older brother, Erik, and flew to San Juan. After taking in the old fort and other sights, we boarded *Adventurer* on Saturday, June 23.

I had worked hard beforehand preparing my six onboard lecturers; back then we used 35-mm slides. For me, at age 34 and likely the youngest so-called expert, this was a kind of a busman's holiday. Early in the cruise, all passengers and lecturers were assembled in, I think, the dining hall.

I had to overcome microphone and stage fright for my overview talk to this large, non-geological audience about what I would cover. Of course, ocean liners are not built to function as schools.

The 'classroom' I was assigned was filled, against the walls, with slot machines, and most of my audience sat on the carpeted floor or stood leaning on the slots. We lecturers often sat in on each others' song and dance. One who literally sat on the floor to hear my plate tectonics spiel was astronaut Russell "Rusty" Schweikart, who had piloted the Lunar Module on Apollo 9. Rusty's lectures were about his experiences in space. I later learned that Neil Armstrong, the first human on the Moon, was on *Canberra*, along with astronaut Scott Carpenter.

The other astronaut we had onboard *Adventurer* was none other than Wally Schirra, a Navy pilot and early astronaut in the Manned Space Flight Program and Project Mercury. A VIP. By 1973 he was retired and his Science at Sea talks were on conservation and energy. My wife Randi recalls Schirra approaching her while she was standing alone along a railing without her husband, who was engrossed in explaining seafloor things to a clutch of other passengers. We lecturers and wives sat at the same tables for meals, and became acquainted. Had he had such a pretty wife, Wally said, he would not have left her alone like that.

Another VIP lecturer with public fame was Arthur C Clarke, who shared an Oscar with Stanley Kubrick for the classic 1969 screenplay *2001: A Space Odyssey*. Clarke projected and commented on examples of space art and predicted what might happen in the 21st century. I should have taken notes.

I figure he predicted we would have international lunar and Martian bases by 2025. Clarke was not exactly a scintillating dinner conversationalist or lecturer, but so what?

As we headed east away from the Caribbean, everyone's attention focused on the real cruise VIP celebrity: the meteorologist. This man would study the weather and climate history and choose the best spot along the eclipse path where we would rendezvous, stay put, and wait for the shadow to arrive. There were no early hurricanes in the area, but even trade-wind clouds could spoil everything. We slowed the ship; location was 11° 30' N, 43° W, Equatorial North Atlantic Ocean.

Well before sunrise on June 30, the decks were turned into a tripod forest loaded with telescopes, cameras, and passengers. The sun rose at 5:30, and the eclipse began at about 6. There was silence on deck, awe, and anticipation as it started to grow dark. With some clouds around, most everyone was hoping a cloud would not drift between us and the Moon. The Sun and Moon stood 18 degrees above the horizon, and no clouds came in the way. Cheers rang out; our weatherman was a hero. Totality lasted a tad more than four minutes.

For two of us, it had all begun with a magazine ad, a long shot, and abundant good luck. About the only folks who were not so thrilled were the bartenders, who complained about these peculiar passengers. They would drink very little and retire early—to get up at dawn to hear a talk on seabirds.

CHAPTER 8

Deep Sea Volcanoes, Basalts, Bite Marks

In the olden days, research ship-port stops meant mailing postcards back to the States. Of course we kidded ocean newbies about supposed mail buoys, much as Boy Scout tenderfeet are taken snipe hunting. There was a lot to write home about on my 1974 investigations near the Galapagos Islands. I was chief scientist researching the plate boundary between the Cocos and Nazca tectonic plates. So I sent two postcards home, the first from Panama, the second one a continuation of the first. As it turned out, my second post card arrived first, and began with "and the teeth marks are still in my arm."

The Galapagos Archipelago is a cluster of geologically young volcanoes, most of which are located on the Nazca plate. Geologically young means a few million years or less. Some of these volcanoes are still active or dormant. Charles Darwin had stopped there on HMS *Beagle,* and his observations on different but similar species of finches and tortoises helped lead him to Evolution of the Species.

Most of us on *De Steiguer* would have loved to visit the islands; we saw the higher ones rising above the horizon off in the distance. But ecotourism was not our mission. However, we had to get Ecuador's permission to research the surrounding ocean, so we stopped at one small desert island called Baltra. Our *De Steiguer* skipper and I, and maybe also a crew member, went ashore. The Ecuadorian military man in uniform who greeted us required some signatures. I can't recall if he had to be bribed as well. His small stucco office and residence were surrounded by a fenced-in area in which resided one medium-large tortoise.

Back then cruises were all male and some in crew, officers and scientists, had girlie magazines like *Playboy* in their cabins. On this expedition someone taped a *Playboy* centerfold to a lab instrument rack. Soon there were copycats. Then Chief Scientist Vogt ruled them inappropriate distractions so they disappeared. Peter the Puritan I was.

Our expedition mission was to drive our vessel, towing a proton magnetometer, repeatedly across the plate boundary to collect profiles of the rough bottom topography and the linear magnetic anomalies paralleling the plate boundary. These, we knew, were the result of seafloor spreading and reversals of the Earth's magnetic field—what turns compass needles. The basalt lava rising up and cooling into rock at the plate boundary would have been magnetized in either of two directions, depending on the polarity of the Earth's magnetic field, which reverses at random intervals every few hundred thousand years on average. In 1974, we geologists already knew the reversal times, so we could use the spacing of the linear magnetic anomalies to measure the history of separation speed between the Cocos and Nazca plates. The oceanic crust, as first recognized in 1963, is a bunch of giant tape recorders. On our cruise we were measuring the speed of one tape recorder.

On each crossing of the plate boundary, we also located its exact location—where the most recent and youngest volcanic rocks were to be found—and where we then intended to dredge up some samples. Basalt lavas erupting on the ocean floor are rapidly chilled by cold seawater and often take on bulbous shapes resembling pillows. These 'pillow basalts' have shiny glassy surfaces because the lava cools too fast to form crystals. (Rapid chilling is also key to making glass.) The thin glassy skin on pillow lavas reacts with seawater and in just a few thousand years is altered chemically and visually turns rusty dull brown.

Sometimes the hot magma within a developing pillow finds a hole in the frozen rocky pillow shell. Lava oozing out forms very small pillows called buds. Toothpaste or putty oozing out of a tube at one point resembles miniature basalt buds. Punch a pinhole in a toothpaste tube and squeeze the tube to get the idea.

Beneath the glassy skins of fresh pillow basalt is basalt with crystals too small to be seen with naked eyes. The magnetization of basalts is carried by crystals of a dark mineral called titanomagnetite (maybe even a few crystals of aptly named magnetite). Titanomagnetite—as suggested by this mineral's name—contains both iron and titanium oxides. It turns out that small crystals are better than large ones in terms of magnetization strength. Thus, ocean floor basalts cause the stronger magnetic anomalies measured by a magnetometer towed behind a research ship.

As suggested by earlier magnetic anomaly data from the Cocos-Nazca plate boundary, the basalt magnetization was even stronger than typical where this boundary approached the Galapagos volcanoes. In a paper I published in *Nature* the previous year, I suggested these basalts might be more magnetic simply because they had more iron and titanium. I gave my idea a catchy name—magnetic tele-chemistry—because if magnetic anomaly strengths measured at the ocean surface were a measure of iron and titanium content, which would likely correlate with other elements among the minerals in those basalts.

Of course, rock magnetization is the result of many other possible factors, so I took some legitimate published flack from another scientist more knowledgeable about rock magnetism. That's how science moves forward. Science flack is not political or personal flack.

The Galapagos volcanoes are not directly on the plate boundary but comprise a so-called hotspot similar to Hawaii or Iceland. (The expert Galapagos volcanologist was Howell Williams, whom I had met as fellow lecturer on the Cunard *Adventurer* in 1973.) Hotspots were postulated by W. Jason Morgan, of Princeton University, in 1971 as expressions of plumes of presumably unusually hot mantle rising from the 1,800 mile-deep base of the Earth's mantle. Down there the Hades-hot mantle was further heated by contact with the molten nickel iron core. The much-analyzed (only by computers, of course) Core-Mantle Boundary is where heat escapes the core and heats the lower mantle. As the mantle materials rise, they begin to melt once within and above 100 miles of the Earth's surface. The mantle does not melt completely, but what does melt rises to the earth's surface and may erupt as lava in volcanoes or along separating plate boundaries.

Hawaii is a case where the plume rises under the middle of a plate (the Pacific plate) while Iceland is the case of plume rising nearly below a spreading plate boundary. The Galapagos hotspot is somewhat in between, but close enough to affect a nearby spreading plate boundary—the one we were investigating on *De Steiguer*. While the plate tectonic paradigm was rapidly accepted by geologists about 1968, the hotspot/

mantle plume hypothesis remains controversial more than half a century later. In particular, there is so far no compelling evidence that mantle plumes, even if they exist, come from near the core-mantle boundary.

The profiles we collected across the plate boundary I hoped would show that mantle material upwelling from the Galapagos plume was flowing from the plume toward and then into the melt strip below the Cocos Nazca Ridge and then east and west along (under) the ridge. Maybe those strong magnetic anomalies and iron-titanium basalts were derived from this plume? The data we collected were interpreted by Dr. Richard Hey, also a Princeton graduate, in terms of propagating rifts. Is this propagation pushed by mantle material flowing east and west under the plate boundary? The jury on that is still out 50 years later.

Once we on *De Steiguer* began to dredge rocks along the plate boundary, our venture became more of an adventure. Dredging is a bit like fishing except we try to snag rocks. Back then we had no camera on the dredge. Just a weak link just above the dredge. In case it anchored the ship, we would sacrifice just the dredge rather than breaking and losing costly cable or damaging the winch. Ideally, the dredge teeth would clip off buds or small pillows, which would then slip back into the chain-link net. We onboard could watch the gauge showing stress on the cable; the dial would go up whenever the dredge snagged a rock, and rapidly come back down once the unseen pillow or bud had broken off or the empty dredge just slipped over the rock. After several such nibbles, or bites, the winch operator would start reeling in our catch, if any. When after an hour or more the dredge finally appeared back at the ocean surface next to the vessel, many eyes would be looking at the haul—or an empty dredge.

One of our dredge hauls included rock types practically never found in oceanic crust. In fact, when we published the analysis—led by a petrologist from the Smithsonian— some skeptics wondered about this having been dropped by an Antarctic iceberg. An iceberg reaching the equator? Not a chance. No IRD (Ice Rafted Detritus) has ever been documented in that area.

The Cocos Nazca Ridge has the distinction of being the site where hydrothermal ecosystems, with their giant clams and tubeworms, were first discovered—by my colleague Kathleen Crane, in 1976, while she was a PhD student at Scripps. The giant clams were remotely photographed from a deep-tow system. Her skeptical colleague Peter Lonsdale knew about our 1974 *De Steiguer* cruise and suggested—perhaps in jest— that the clam shells had been tossed from *De Steiguer* after a shipboard clambake. Kathy later named this first site Clambake, and for many years I thought that name related to the hot water emanations there. Could we have dredged up such clam shells and discovered the first chemosynthetic ecosystems on *De Steiguer*? Of course, but we didn't. No way now to find out how close our dredges ever came to such a vent. Maybe I don't want to know.

On one of our last dredging stations, we got the vessel stuck, anchored in the open ocean. We tried to get the dredge free by steaming in all four compass directions. I was up for 24 hours. Finally I gave the approval for all engines go. The weak link would break and we would leave this dredge on the ocean floor. I couldn't bear that experience, so I disappeared into my quarters and fell soundly fast asleep in my bunk. I woke up late and looked down from my bunk. On the floor was a huge pile of shiny black pillow basalt. To some, it might resemble a bowel movement by Vulcan's oceanic cousin. To me, it was a pile of gold. My team had carried the rocks in from a filled dredge, but without waking me.

Now about those teeth marks. They were not left by some shark. Or sea serpent. Or a Panamanian dog. On our Panama port stop (for food, fuel, etc), I was walking down along a busy seaside avenue on a typical hot day, watching vultures circle high above the city dump. Suddenly, from behind, human arms were around my neck: I was being robbed. I tried to hold on to my wallet.

The second assailant grabbed and bit me just below the elbow. So I let go of the wallet. Those tooth marks were human! Robbers evidently used teeth rather than guns or knives because the Panamanian dictator, Omar Torrijos, showed armed robbers no mercy. Even knives were verboten. Or so I learned. The robbers darted across the avenue, dodging traffic.

I flagged down a police jeep, but they showed no interest, let alone sympathy. To them I was probably one of those despicable Canalistas. Our voyage predated the United States turning the canal and over to Panama. And 1974 also predated AIDS/HIV. Just prior to departing Panama, I wrote home a letter that again mentioned the attack. I mentioned that my arm was still sore, and so was my neck, the latter due to the first robber grabbing me around my neck before his accomplice bit.

This was only the second time I got bitten by another human. The first time was back at Caltech on a drive-in movie date with the sister of a physics major classmate. She insisted on smoking in my old car, so I reached over, snatched the cigarette out of her mouth and tossed it out of the window. The bite marks were in my right forearm but did not last long. It was the first and last date with her. I did not have much of a social life then.

Letters Home

1974 De Steiguer cruise, eastern Equatorial Pacific: Get off and on San Diego, port stop Panama, landing party on Baltra Island, Galapagos to get permission from Ecuador military to work in Galapagos waters. We had some Ecuadorian Lt and his wife on board for that.

First part of a letter home mailed soon after 16 January by an Ecuadorian Lt who visited De Steiguer from Baltra in a landing boat. Some of us had previously gone ashore briefly—a small desert island with a lone tortoise living in a stone enclosure, and a shabby stucco barracks.

January 10, 1974, Evening
Ecuador, Galapagos 99° W, 14° N

Dear far-away but close-by wife!

Despite my mild bout of dysentery, the tropical
ocean outside is so stunning it tickled my poetic
bone, and what better time to write. Inside this
ship, named for Admiral de Steiguer, Hydrogra-
pher of the Navy in the '20s, it looks no different
than the *Bartlett* or the *Lynch*. Differences are
minor. Only the faces are all different, all but one
or two I hardly knew, who were on the *Bartlett's*
crew two years ago. Sitting in my 'stateroom,'
the same one I had on *Lynch* (and you saw), and
(with a certain Bracey) on *Bartlett*, time seems
to have stopped. A kind of reverse culture shock,
a kind of daze. The whine of air conditioners,
the engine vibrations, hammering, and scraping
somewhere, as always.

My GOFAR crew is a competent one though,
and dedicated to make this a model one in terms
of quality of data, the way it is marked and folded,
etc., and the quality of navigation. Hemler,
Lanassa, Madosik, and young long-haired mild
mannered David Clark of the Gravity Division.
Yes, this time we have a working gravity meter

on board! Even the AGOR coordinator and the
ET are easier to get on with than their counter-
parts—priggish and gruffish, respectively—on
the *Bartlett.*

Carey Ingram is the coordinator, casual, informal,
almost but not quite to the point of being disor-
ganized, and big, like Ed Escowitz. Cary Ingram
was on Fred's [Fred Bowles, a sedimentologist
GOFAR colleague] cruise down to these parts,
knows Kravitz, Glover, Doris and Jim Rucker,
etc, and many of the mostly NAVOCEANO
people on the *Northwind* 1965 icebreaker
cruise to Soviet Arctic, and he was well aware of
the nasty feud between the *Atka* Exec and the
NAVOCEANO people on board. I told him
you and I got engaged in Bergen and that Capt.
Blake of the *Atka* had sent us Wisconsin people
such a fine letter.

Ingram likes to tool around the California
desert in his Dodge pickup, and his wife
promised him one day out in the desert for
every 5 lbs he loses on this cruise. Right now
our coordinator is building a home-made
model steam engine (!) out of scrounged parts,
which is something Escowitz would do. (Ed
Escowitz, GOFAR coworker not on board but
with me on 1970 *Lynch* cruise.)

The ET is a short, also overweight Latin type,
good but not excellent, but persistent. Quite a

pile of blubber when he, Lou Hemler, and Ingram are all wedged in there between our various recorders. And all puffing on cigarettes. All with a statistical life expectancy of 50, I imagine. Which leads me to ponder on the death last week of Billy Moore's counterpart down in the Maury Center, Jim Gassaway, whom I knew rather well and spent an hour talking to the last time I drove down to NRL. Not fat and not a smoker and not yet 40 or barely so. Herb mentioned it when he called me on the ship prior to departure.

Leaving you, leaving those sparkling kids of ours, leaving the holiday cheer of my parents' house in Santa Barbara [about 4 hrs drive north of the ship tied up in San Diego], then a catacomb of gray and noisy steel, well you can imagine the death of a close acquaintance is just the ticket to gloom. Say what you will about your family history [Randi's parents both died in their 50s] and mine, I am a hard-working male 5 years your senior, and my South American cousin Kurt, also robust and hard-working, didn't make it past his mid-40s. I don't want to depress you but communicate my thoughts to you I must...

Well on the bright side, we passed near all 20 of the shallow spots assigned and didn't run aground. Nothing less than 32 fathoms, in fact. (On an opportunity basis, i.e. more or less en route, AGOR ships were to verify or—most likely show non-existent—reported shoals—i.e., possible hazards to navigation—still printed on hydrographic charts.)

Also today, finally after a frustrating week, we are getting some reasonable seismic reflection records. Mind you, we are still en route, plowing along day after day. We have left the cool waters of the California Current behind us, evident even a day south of Cape San Lucas. It took us two or three days to run out of the influence of that great rainstorm that must have lasted until you left S.B. And perhaps even snow in Santa Ynez? It was cloudy and sprinkly when we left San Diego, and we had a bit of rain down to the desert latitudes of north-central Baja.

Some of our shoal spots were within sight of that stark and somber landscape of Baja California. Our last landfall was just north of Cape San Lucas, where from 20 miles at sea the great Sierra de la Giganta (6500 feet) forms the skyline. I spent some time on the bridge trying to help identify landmarks, mountain tops to correlate with the radar picture and what was sketched on the charts. This to get fixes, since we had no satellites for hours. And then the full Moon rising over the mountains. And the only sign of civilization at intervals is a faint blink caused by a solitary car winding around an invisible mountain road, his headlights occasionally swinging out to sea.

A few moments like these are the only time it seems worthwhile when, briefly, the cruise has some soul. There are a few other moments, like scientific discovery of something, that may still lie ahead, since we are still not where we want to be.

January 16, 1974

Dear Randi,

This has to be quick—tell you all later. Lots of love! Later letters will have more questions of how life is treating you and the boys but then I'll have some mail from you.

The lieutenant and his wife are leaving now in the landing boat, they'll mail lots of love again from the cactus lava island of Baltra—5 sailors, 7 airmen, 1 woman (the lieutenant's wife) and 1 turtle. But our short stay was great, see you later. Take Erik out to see the comet if you haven't.

January 28, 1974

Dear Randi,

A late rainy evening about 4° 30' N, 89° 30' W, en route to Panama and that long-awaited thrill, WORD FROM HOME. I feel locked in a grey cage, a floating frustration. Be with me in this gloom of separation. It seems so long since January 3rd—long enough that I should by rights be in your arms again. But no way; we have another leg to endure, but that will be the home stretch. Oh to be home and stretch out in your arms. And my little bambinos, goddamn if I don't miss the hell out of them! Maybe Jason will be trained and talking a word (not a streak) when I return. By the time you read this it will be PAST HALF over. You know how to make the time pass with less torment, keep yourself busy, work hard, plow into those projects, it works for me. I have been, with my jolly team's assistance, plotting all our data as we go, with pencil, graph paper, and dividers. It can be done without a computer. It just takes time and constant application. Every morning I am 8 hrs behind again, and the data keeps rolling in. Happily all our major bugs were out before we hit the ops area, so we did well. There are some intriguing stories here, and I have every reason to be satisfied with the science. Hope the next leg goes equally well.

Wait: I just had a vision of you waiting on the dock in Panama. Will you be there? Oh I see you again on the dock in Bergen as we push off for the Arctic. I know you will be waiting for me—surprise—in Panama. See, my stream of consciousness has turned into a raging whitewater of hallucinations.

Despite the good outcome of this leg— surprising considering all the portents and vicissitudes that came before—a cloud of gloom settles over me whenever I am not hard at work. Perhaps I should go to the movies—this evening I saw the excellent one about an Italian soldier (Marcello Mastroianni) saved from freezing on the Russian front by a Russian girl whom he marries, while his Italian wife (Sophia Loren) chases after him. Sunflower! An excellent, sad, romantic, tragic flick.

Usually I work during movie time anyway. The goddamn taxpayer who pays my overtime gets his f*** money's worth!

[NAVOCEANO paid overtime for Saturdays and Sundays, while NRL where I worked after 1975 did not pay scientists overtime, but only staff and technicians.]

Maybe I should work on an art object. I have spotted some brass stock in the engine room but still wait for time and inspiration to take a hacksaw to it. Or I could learn to play poker, the hot thing on this boat. Almost every evening, sometimes till just before breakfast, a motley and integrated circle of cardsharps collect around the rec room table, smoking and talking about winning or losing 10s (or 100s?) of dollars. The circle includes an AB, the coordinator, our young gravity boy, the captain, the radio operator, a room steward, and others.

I guess I'm not of that ilk. Rather brood out over the trackless waves, I guess. Never be any different. I had a slight run-in with the captain this morning. It seems his pet peeve is SNRs doing last minute maneuvers with the ship.

[SNR means Senior NAVOCEANO Representative, their equivalent of a chief scientist. GOFAR was part of NAVOCEANO but used chief scientists because we were doing science not just mapping.]

Would I make us late for our in-ports? Must not keep the pilot waiting, you know. He heard me call a course change from 164 to 173 which he thought would take us away from Panama. He was fuming in his Bermuda shorts in the wheelhouse. Well, I was firm and he was wrong, a false alarm. It was merely to get us back on a course we had already agreed on. Well, everything worked out ok and in fact we are well ahead of time, so he can sleep in peace, the SOB. With that exception, relations on all sides have been excellent and in fact last week the captain told Ingram that our group was as agreeable to work with and as well organized as any, especially the NAVOCEANO groups who use the ship. Apparently they get on board and boss everyone around and treat the captain like a cab driver.

Enough for now, my love. You check that our bambinos are tucked in and then put down your book and go to sleep. Hope I will dream of you standing on the dock in Panama!

January 30, 1974

Dear Randi,

It's night time with a hazy moon, light airs, and swell from the south. Crossing the Coiba Ridge just south of eastern Panama. Lots of drifting logs and stumps all day, perhaps floated off the beaches by that ultra-high tide two weeks ago. We pulled our maggie and sparker cables in to avoid a chance accident. At least one log per 2000 feet of ocean adds up to a lot of flotsam. Some had boobies riding on them. One log turned out to be a SEA TURTLE. There, I finally saw one. And a sea snake. And we hauled in a Japanese fishing buoy and floats, adrift. A head of a swordfish was left on the hook, a la *Old Man and the Sea*. Shucks, the Japs use plastic balls now, instead of the old glass floats. Our resident fisherman also hauled in two wahoos (3 ft long, like barracudas) and two mahi mahi (dolphin fish), one of which I fileted. Now there will be fresh fish in the galley! Passed a Japanese fishing boat this evening, first ship in weeks at sea. My laundry should be finished by now, and it's late. Think loving thoughts of me.

February 4, 1974

Dear Randi,

I've moved back on board *De Steiguer*, and there is time for some more words from my spirit to yours. It's just past 8, and we don't sail until tomorrow morning, a good night's rest away. My arm is still sore with obvious teeth marks where one of the assailants bit me (I hope the miserable bastard doesn't have rabies), and my neck also, where the other one had me...Oh well...Please do work on notifying everyone about the losses.

A bright moon shines outside, the clouds are almost gone. The air is rich and warm and full of secret fragrances of jungle plants. We're tied up at a small and very quiet naval facility on the opposite side of the canal from Panama City, and just a bit inside the canal itself from the Gulf of Panama. At sunset I walked up to the great suspension bridge, I think only a decade old, that spans the Canal. Armed with new binoculars and a transistor, I enjoyed a stiff breeze and the stately magnificence of great ships going in and out of the gateway between two oceans. A freighter from Hamburg was one of the four that passed beneath...

Of course if you view the Americas as essentially north-south you will be terribly disoriented—there is a place where the sun rises in the Pacific and sets in the Atlantic. Flocks of great white birds flew under the bridge too, from the Pacific, flying up the canal. There was a sense of purpose in their flight, as with...

[Rest of page missing]

You're a wonderful correspondent, perhaps a more personal one than I, but part of that may be the new scenes and settings that you want to depict for me...

[Many lines of personal mush omitted]

I dream of dense darkly forested tropical mountains, the soft swash of the tropical sea, the distant chicken or a dog, a musk of strange flowers and the sea, a bit of coconut and a hazy smoke of brush burned to make room for banana trees, brilliant stars of the southern sky, Orion and Sirius overhead instead of low on the evening horizon, dream of the good times we had on the Cunard *Adventurer*, Eclipse cruise 1973, on which I lectured the fjord-like majesty of St. Lucia, the noise and bustle of Trinidad, the hot hot sun and the blue sky and the salty sea of St Thomas.

February 4, 1974
Postcard from Panama

Dear Beloved,

Tomorrow we leave Panama. Good, I've had enough. Think of Port of Spain, Trinidad, visited on the 1973 *Adventurer* cruise with you, but noisier, dirtier, larger, and the Canal Zone Problem beclouding things. I've learned a lot—even visited some whore houses to get the total scene. If you were here maybe you would have wanted to take a peek at these places but they say the girls here have beat up Gringo girls escorted in.

You married a man much too romantic and in love to be tempted even after 28 days at sea! I did drink a lot of beer though and alas got attacked and robbed of my wallet in broad daylight—after all these years. Well, I'm alive and well, mainly pissed off. Please inquire about the new driver's license & ... [Continued on glossy front of card but hard to make out.] Mainly loss of car registration card and a blank check.

CHAPTER 9

Drilling Deep on
DV Glomar Challenger

Glomar Challenger

Glomar Challenger was a deep-sea drilling vessel built in Texas for oceanography and marine geology studies. *Glomar* is a truncation of *Global Marine*. *Challenger* is a tribute to the 19th century oceanographic vessel HMS *Challenger*.

Hands down, my 1975 stint as co-chief scientist on the DV *Glomar Challenger* was the most rewarding expedition in my career.

DV stands for Drilling Vessel. This particular ship was converted in 1968 from commercial oil and gas exploration to ocean floor geoscience in the Deep Sea Drilling Project, sponsored by the National Science Foundation. At 400 feet long and a wide 65 feet, this legendary vessel displaced 18,600 tons, the heaviest ship of my ocean geoscience career. The derrick towered 140 feet over the sea, supporting a drill string up to 22,500 feet long. That's about four miles of pipe hanging down from the ship into seafloor sediments. *Glomar Challenger* had enough fuel and supplies for a crew of 25 and up to 25 scientists to remain at sea for 90 days.

I use the past tense because she was retired and scrapped in 1983 and replaced by the bigger and more advanced drill ship DV *JOIDES Resolution.* But back in the '70s, there often was confusion with the *Glomar Explorer*, designed to covertly retrieve the wreck of a Soviet submarine near Hawaii while pretending to search for commercially viable manganese nodules on the ocean floor.

Drilling legs take many months of planning and scientifically well-supported proposals. These are vetted by several panels and address not only research merits but also technical feasibility and safety.

Deep-drilling legs always have a pair of chief scientists, and my co-chief scientist was Brian Tucholke, a marine geologist at the Woods Hole Oceanographic Institution. I had spent seven months on one of their ships, RV *Chain*, as a watchstander a decade earlier.

Brian and I each chose several drill sites. All sites were of some interest to both of us. Brian was most interested in the sedimentation and paleo-oceanographic history of the western central Atlantic away from the North American margin and its wedge of sediments eroded from land and dumped into the Atlantic. The subsequent Deep Sea Drilling Project leg 44 would focus on that debris wedge. Brian wanted cores from the Bermuda Rise because it remained above and free of such continental detritus.

Brian and I were both interested in coring the sediments on the crest of the Bermuda Rise in an old fracture zone valley filled with thick sediments. The site we chose was not far from the Bermuda volcano on whose beveled summit the island sits. Past drilling on Bermuda penetrated its carbonate cap, but basalt samples below that cap could not reliably date the time the Bermuda volcano grew from the ocean floor and above sea level. Maybe *Glomar Challenger* could recover volcanic debris and the time it erupted and arrived on the ocean floor.

For my part I hoped to get rock fragments from two of the New England Seamounts. Getting ages from those fragments would help test the idea that this seamount chain resembles the Hawaiian chain with its progressive age increase from the young active volcanoes on the Big Island westwards. This age progression had been attributed to a fixed hotspot or mantle plume across which the Pacific tectonic plate was moving. This is often likened to charred paper as a sheet is moved across and above a candle flame.

My second drilling target was a mysterious ridge in the oceanic crust. Most of the ridge was buried below the sedimentary cover. Associated with this J-Anomaly Ridge was a high amplitude magnetic anomaly measured by ships and aircraft. I wanted us to drill through (and sample) the sediment cover into the likely volcanic oceanic crust. We would get at least some rock samples from that ridge. Would those rock samples be highly magnetized basalts? Or were the magnetic sources deeper?

Our all-male science team flew from wherever they were to Ponta Delgada in the Azores to meet each other, mostly for the first time in person, and board the drill ship, which had arrived from Istanbul at the end of the previous leg. Our team was international and included specialists from South Korea, the UK, and Germany. Also on board was the Soviet-Estonian sediment mineralogist Ivar Murdmaa, who is featured in the following chapter. Nations that helped finance the project got to place their scientists on the ship and participate in data analysis and publication.

En route westward to our first drill site on Nashville Seamount, we had time to get acquainted with the ship and each other. The *Glomar Challenger* was a floating laboratory where I would be surrounded by experts in various disciplines. The team was assembled for this leg, and specialists assigned to the vessel by Scripps Institution of Oceanography, which operated the vessel.

The voyage was so much more than being at sea or in the air just to measure magnetic and gravity fields. Leg 43 coreholes ranged from okay to total surprises, the most exciting one thanks to winning the toss of a quarter. Let's save that story to the end.

We tried to core two seamounts, but as usual had trouble spudding the drill into the loose volcanic talus (rubble). A datable fragment was returned at one site, and the plankton mixed with volcanic ash gave an eruption age because the species of plankton lived during a relatively short period of time of known age.

A similar method allowed us at Site 386 near Bermuda to determine the time the Bermuda volcano had grown above sea level and was erupting, about 45 to 50 million years ago. The next site to the west returned, by a different method, a similar age for the wide Bermuda Rise.

Site 383 drilling was planned to reach the buried crest of the mysterious J-Anomaly Ridge on the flat Sohm Abyssal Plain. However we failed to penetrate far into the thick sands deposited on the Abyssal Plain by numerous avalanche-like turbidites, also known as suspension flows, mostly during the last ice age. These flows began on the Canadian continental margin near the Grand Banks. A wide fast-flowing glacier once drained the great Laurentide Ice Sheet, and sediment eroded from present Canada was carried both by the glacier and by streams flowing under the glacier.

Although the ice sheet is long gone, the glacial sediments brought to the shelf edge continue to fail. They are often triggered by earthquakes. One such turbidity flow broke several seafloor telegraph cables in succession. The cable break times were recorded and allowed well-known marine geologists to estimate rapid flow speeds. Commotion in the ocean!

Once we agreed to abandon the site, it would take several hours to pull up the drill string, 30-foot sections at a time. Scientists use the metric system but the drilling was inherited from industry, which then and likely forever uses the ancient English system. Our Navy similarly uses yards and kilo yards but that makes some sense: 1,000 yards is a nautical mile, or one arc-minute of latitude.

As the drill string was coming up, the ship's drilling superintendent asked Brian and me about our next site. Because 283 was the northernmost planned site, Brian was anxious to steam south toward the Bermuda Rise. However I pointed out that the J-Anomaly Ridge rose above those terrible sands not far to the north. We had seismic reflection profiles of the ridge showing acoustically transparent, easy-to-core hemipelagic sediments capping the ridge crest.

I wanted to go north and Brian insisted on going south. Neither of us would budge. We had a civil, scientific-logistic disagreement. So we decided to assemble the entire scientist team, where each of us presented our case. We left the room so the team could debate in our absence. Bob Houghton, a young igneous-rock grad student, acted as courier and abstained.

The science team was divided 50–50. The last sections of drill string were just coming up. What to do? The drilling superintendent was probably cursing the scientists. I suggested we flip a coin. I won. DV *Glomar Challenger* prepared to turn north to Site 384. Soon after the first sections came up, I suspect Brian no longer regretted losing the coin toss. But we never discussed it to this day.

Before I describe what Site 384 revealed I want to get back to Bob Houghton. He was bright and personable, a grad student at MIT doing lab work there and research at Woods Hole Oceanographic Institution.

His thesis involved the New England Seamounts and their ages and compositions. After we got back, he wrote the chapter on the New England Seamounts, including the two we tried to core, and various other data, some of which he attributed to

Bob Duncan at Princeton. Houghton's paper, which supported a Hawaiian-like age progression, was published in the 1979 *Initial Reports*.

It turned out that Houghton had simply made up lots of data. Duncan never made those rock-age measurements. I recall Houghton was booted out of MIT and Woods Hole, probably very embarrassing for both prestigious institutions. I don't know what happened to him.

On our watch, we chief scientists donned yellow safety helmets whenever another 30-foot core came up and was laid on the deck. As we drilled farther into the soft sediments capping the J-Anomaly Ridge, the sediments naturally got older and older. We would take a spatula of sediment from the core catcher and race down several flights of stairs to the core lab. I would hand our Korean micropaleontologist the spatula and he would quickly smear some on a slide to look for certain diagnostic foraminifera. He would note the specific stratigraphic zone, inform me of his findings and I would race back up and write that zone and age on a chalkboard. We were well down into the early Danian epoch, a time very poorly known from deep-sea cores.

Finally, we cored through the Cretaceous-Tertiary boundary. The lower Danian sediments were darker in color than the underlying yellowish Maastrichtian mud. The time of dinosaur extinction was missing; sediments from that asteroid impact had not then been identified. In fact evidence for an asteroid would not be discovered for several more years.

The break in sediment type in Section No. 13 could simply have reflected lots of missing time but not a sudden catastrophe. I discussed this with the shipboard paleontologist. I said maybe it was a catastrophe, but he smiled and accused me of long-outdated Biblical thinking. Core No. 13 had come closer than any before to capturing the extinction.

Years later, the length of time missing was estimated at 0.9 million years. I suspect megatsunamis caused by the asteroid impact washed the missing sediment away.

The drilling proceeded further into the Cretaceous, and our seismic reflection profiles suggested we were reaching the basaltic crust of the J-Anomaly Ridge. Indeed, the ship shuddered a bit. I went out onto the drilling platform. The driller was looking at his gauges.

"That doesn't feel like basalt," he said.

He had drilled into basaltic oceanic crust and he was right. When that core section came up, Brian and I were surprised to see coralline sand in the core catcher. We had cored into the deepest subsided former island known in today's oceans.

From reef fossils, we would know that we had drilled into the 120-million-year-old early Mid-Atlantic Ridge. We also cored some actual basalt but it was too altered by seawater to measure its age. And what about the high magnetic anomaly? We didn't solve that mystery—but it's good to leave some mysteries for geologists of the future.

Letters Home

25 June 1975
Ponta Delgada; postcard

Dear Randi,

Slightly dazed after a short and sleepless night, I've now met all the participants & we're all in the same hotel, overlooking the waterfront promenade and harbor. Rothe the German arrived at noon; McCabe the Englishman also, while Murdmaa the Russian, who we thought might not make it, has been here 3 days. Unfortunately the ship won't arrive until Friday noon (the day after tomorrow). It's mild and marine here, like Santa Barbara & the Washington heat (which extended to Boston!) seems far away. Have a good July (4 July is our first hole) and think often of your husband/father.

The airport incident proves again that I'm helpless without you.

9 July 1975
Western Union Telegram

Port changed Norfolk ETA 13 August. Hope July going well and Erik and Jason being good. Pulling out of first hole. Work challenging, weather excellent except tropical storm Amy. Loving thoughts, Peter.

22 July 1975

Dear Randi,

Just tied up at Hamilton, Bermuda due to an unscheduled port stop for repair of bow thruster.

Surprise! Bet you didn't expect a letter. It's an ill wind, etc. After our last site (#4, including a total abortion, where the abyssal plain turbidite sand continuously caved into the hole and we had to pull out again after only 100 m penetration) a salt water leak...

> *Letter continued on a bench in a fort-moat*
> *turned garden in late afternoon sun.*

With various drunken shipmates I made the rounds of all the bars until each of them closed, there's always one more still open just down the street. We hit the last bar about 3:30. Yecch do I feel horrible! At one point the young 3rd Mate & I took in a Trinidad-Bermuda steel drums band that played, believe it or not, excerpts from The Messiah, Poet & Peasant, & even Johann Strauss is alive and well among the 'pans.'

> *Letter continued late the next morning*
> *in my cabin—with a horrible hangover.*

The reason I didn't finish this letter in the moat garden among the birds and flowers is I realized it was 5 p.m. I had read they closed up at 5. So I walked around to the exit tunnel which was dark because the lights were turned off and then I came to a locked door. Yes I had been locked in the moat! All shouting was to no avail & couldn't find any exits. Finally & desperately I climbed up the side of the bridge out of the moat & so 'saved' myself. The way I feel now I would have been much better off spending the night in the moat. It looks like the bow thruster will be repaired by afternoon so I should get this letter mailed. It seems like a major effort but for you it's worth it. A few more weeks and I'll be home again & the two trips will be behind us...

This *Challenger* cruise is the most rewarding and exciting of any cruise, with no close competitors. It's also the hardest work. There's scarcely time to sunbathe or take in a movie! The shipboard scientists have to put together a very thorough looseleaf book summary of the findings, so there's that to work on when not supervising the drilling. When I get back I can tell you the strange story of Drill Site 384 in detail, but consider for a start core #13, retrieved on the 13th of July, on a cruise that sailed for the Bermuda Triangle on Friday the 13th and was to return on the 13th.

Today, it's known from other sites that an innocuous bunch of creamy brown foraminiferal-nannoplankton ooze continuously covers the Cretaceous-Tertiary transition. (The end of the age of dinosaurs and 75% of species today, called the Cretaceous-Paleogene boundary and dated to about 66.0 million years ago and attributed largely or entirely to an asteroid impact.)

You may have heard that the last page of the Cretaceous chapter and the lowermost Tertiary chapter seems to have been torn out of the history book of the ocean floor—or else it was not printed at all! After 7 years of drilling and on the last DSDP leg, we finally found a spot (more or less by luck, and in the final analysis because of my winning a flip of the coin) where paleontologists might read the oceanic scroll nearly to the last 'days' of the dinosaurs and all that died off when they did. What clues our cores might hold about the nature of the 'Time of Great Dying' who can say now? More about Station 384 later.

[In later years the actual Boundary Clay, asteroid fallout ca ½ to 1" thick layer, was found at several sites around the world, and our core did not record the entire complete transition, but came as close as any at the time. Enormous tsunamis triggered by the impact, and subsequent very low sedimentation rates, probably explain why the transition was not recorded at earlier drill sites.]

Telegram from Randi

RCA Global Telegram

Sent to PETER VOGT

MV   GLOMAR CHALLENGER

HIGH SEAS SITE #5

LAT 31° 11′ 23″ N, 64° 14′ 69″ W

LOVE FROM US. PROPOSED EASTERN
VACATION FROM NORFOLK INSTEAD
US   CROSSING   FLY   CALIFORNIA
CHRISTMAS.   OPINION   PLEASE
LOOKING FORWARD TO NORFOLK AND
YOU ALL WELL HERE

LOVE RANDI

RCVD AT 24 JULY 1550Z

CHAPTER 10

The Vela Incident

Adventures with challenges and adrenaline rushes can also happen back in the office. This story is about the so-called Vela Incident on the early morning of September 22, 1979, and its aftermath.

The backstory began with the Limited Nuclear Test Ban Treaty, whose signatory nations agreed to stop atmospheric nuclear bomb testing. To ensure compliance, the US launched certain satellites to monitor cheating. The law Congress passed restricted US exports and otherwise penalized non-signatory nations caught conducting clandestine tests. Nuclear explosions in the atmosphere were known to create two close-spaced flashes of light. Vela Hotel satellites were outfitted with so-called bangmeters to detect such flashes and report any detections to ground receivers.

By 1979, new advanced types of these satellites were planned but not yet in orbit. However, only one of the two Vela satellites monitoring the southern hemisphere oceans was still in operation at that time. A double flash needed to be recorded on both satellites for the flash location to be obtained with reasonable accuracy.

A double flash was recorded by US Vela 6911. The range of this satellite covered a broad area extending from the southern South Atlantic east to the southwestern Indian Ocean. Jimmy Carter was president at that time, and his science advisor, Frank Press, had been a professor of mine two decades earlier. After the event was made public, the Defense Department declared that it was either a bomb blast or a combination of natural phenomena, such as lightning, a meteor, or a glint from the sun.

Press appointed an eight-person panel to investigate the Vela Incident and decide whether it was actually a nuclear bomb test or some natural event. If it was a bomb test, the administration would have to investigate and penalize the nation conducting the test. Suspicions immediately focused on Israel and South Africa, both known to be involved in nuclear bomb research. Press appointed Jack Ruina, like Press an MIT alumnus, to chair the panel, which became known as the Ruina Panel. The panel included superbrain Richard Garwin of IBM and former Manhattan Project physicist Luis Alvarez (who, with his geologist son Walter, had recently first found smoking-gun evidence that the dinosaur extinction catastrophe 66 million years ago was due to one or more asteroid impacts).

After reviewing the evidence they could find at the time, the Ruina Panel concluded that the double-flash was likely a "zoo event" of natural origin, such as glints from a nearby meteor. However, skeptics suspected the panel's conclusion was politically motivated to relieve the Carter Administration from any follow ups.

One skeptic was physicist Alan Berman, Director of Research at the Naval Research Laboratory, where I worked. Berman appointed a panel of about 30 experts to investigate the Vela Incident with data—especially hydroacoustic data from US Navy sea-floor sensors—the Ruina Panel had overlooked. Berman chose me as a panel member, which explains my involvement. I was asked to independently review seismic data, possible recordings by land seismometers.

Berman's panel dropped what we were doing and went to work. We did not have the seismometers at NRL, so I was tasked to search elsewhere. October and November, 1979, were exciting times for me and other team members. All of us on the panel were instructed to stick to the science. Speculation, written or verbal, about which country was responsible or about possible internal US politics was strictly prohibited.

(I should digress to express my thanks to Dr. Berman for approving my proposal to spend a year, 1977–'78, as visiting scientist at the University of Oslo, Norway. Given that I had only been in my job for two years, the Lab's powerful Research Advisory Committee had rejected my proposal. But Berman overruled their verdict.)

Before going further with this story, I should mention my main source: a day-long conference on November 22, 2019, at the Woodrow Wilson Center in Washington. The conference assembled those scientists still alive and involved in the Vela Incident. The goal was to collect oral histories—40 years later. Alan Berman was too ill to attend (he died in 2022) but submitted extensive notes. In his notes, he emphasized that several seafloor hydrophones— notably including those off Ascension Island—had recorded the signals from the explosion and localized them to South Africa's Prince Edward and Marion islands. Moreover, elevated Iodine 131, a short-lived isotope created by a nuclear explosion, was found in the thyroid glands of Australian sheep slaughtered in the two months after the

Vela Incident. This decay isotope would have been carried east to Australia by Roaring Forties westerlies.

The Naval Research Laboratory report—much of it still classified today—was not released in final form to the cleared authorities until June, 1980. A preliminary report of findings was presented to the Ruina Panel at its last meeting, in December 1979. They were not pleased and questioned some of our findings. However, President Carter, who closely followed the Vela Incident investigations, wrote in his diary that it was likely a bomb had been tested by Israel, with the involvement of South Africa.

As one of the few from the Berman team to attend the Ruina Panel meeting at the Stanford Institute, I was more nervous about my presentation than I was before my PhD prelim exams more than a dozen years earlier. A representative from the Defense Intelligence Agency spoke just before coffee break. As I recall, he talked statistics and probabilities. While writing some equations on the blackboard, he was interrupted by superbrain Garwin, who went to the blackboard and finished the equation, explaining afterward that it was all wrong. The Defense Intel rep slunk back to his chair humiliated.

We from the Naval Research Lab stood around outside the meeting room during coffee break; my talk would be next. Berman said if anyone from the Lab did what we had just seen, he would be out the door the next day. Berman was not referring to Garwin's rudeness. Rather it was the Defense Department man's technical incompetence.

I was shaking, at least inside, as we walked back in. But I had done my homework and reported just the facts. Seismology was not controversial here because no signals were identified. Anyway I was reporting on work done by others. There were no questions. Whew!

The mystery around the Vela Incident remains. However, some Israelis, likely from their DC embassy, asked and were allowed to visit the Naval Research Laboratory. They were seemingly saying to us: "Show us what you got."

When my turn came to brief them, I told them all about plate tectonics. I noticed at least one Israeli rolling his eyes.

CHAPTER 11

Prince Andrew and Green Turtles

I was buckled in as the Naval Research Laboratory's Navy P-3 Orion turboprop rolled down the Naval Air Station Patuxent River runway on our long flight to Ascension Island. Led by Lab colleague John Brozena, our 1984 mission was to measure gravity anomalies in a large ocean expanse around the island.

Those P-3 Orions were and perhaps still are ideal because of their long range at low elevations. Flying low gets one closer to the sources of gravity and magnetic anomalies, which attenuate with distance. (The US Navy today uses its Boeing P-8 Poseidons to search for enemy submarines.)

P-3 Orion

I volunteered to come along as an assistant. Yes, partly for the research experience, adventure, and, of course, a visit to the remote, exotic, and geologically interesting Ascension island.

Brozena had developed an airborne gravimeter system, no small feat given the relatively tiny perturbations of gravity due to seafloor mountains and valleys. Such a gravity anomaly would make an average adult weigh, more or less than by the weight of a teaspoon of water. A seafloor mountain, or dense rock mass below the seafloor, is a mass that pulls on you a tiny but measurable bit. What you weigh is mostly how much Earth as a whole pulls on your mass. On the Moon, for example, you would weigh a lot less—as we remember from Apollo astronauts loaded with packs leaping around. Remember Isaac Newton, who figured all this out already in his 1687 opus. A seafloor mountain has one advantage over Earth: The mountain is much closer to you than most of the rest of our planet. However, Earth wins by its much, much greater mass.

A bumpy flight can cause your weight to change by many pounds. Measuring natural (geological) gravity anomalies became possible by the new Global Positioning Satellites, that GPS we take for granted in our cars and smartphones these days. The other thing making aerogravity possible is that the bumpy air and other airplane vibrations are of short duration and can largely be filtered out by onboard computers.

As the ocean passed endlessly below us, I recalled viewing that isolated island of Ascension on my 1975 voyage on the Soviet *Akademik Kurchatov*. It was early evening and from just a few miles offshore, my Soviet geologist colleagues and I had stood along a railing and watched the bright lights of the NASA tracking station (and various US and UK military installations). These mostly Russian geologists had hoped to go ashore and collect some of the unusual igneous/volcanic rocks exposed on this 34 square mile island, 1,000 miles from Africa and 1,400 miles from South America. Not surprisingly, the island was officially off limits. But they naively thought I had some influence to get them permission. Ascension was also rather inaccessible to tourists in the early '80s.

Ascension Island lies about 80 miles south of the equator and 62 miles west of the Mid-Atlantic Ridge rift valley, along which the South American and African tectonic plates are separating at an inch or two per year. Ascension began as a seafloor volcano (seamount), but by a million years ago had grown to rise out of the sea. The Ascension volcano or volcanoes have been dormant in recent centuries. Overall, it's a stratovolcano, geologically not unlike those we climbed in Mexico and Ecuador. The Ascension terrain at lower elevations is largely a nearly barren clinker desert (only seven inches of rain annually, probably less around the coasts).

We would, when time permitted, hike through a landscape peppered with numerous volcanic cinder cones and black lava flows. Enough rain and condensation from nearly perpetual clouds support much vegetation, almost none native, at the higher elevations of Green Mountain (summit elevation 2,815 feet).

Those times we from the Naval Research Lab had free time were not provided for our geological or tourist relaxation and enjoyment but were mandated to give our VXN-8 squadron pilots enough rest time between flights. And to enjoy a beer or cocktail at the Royal Air Force base O-Club—but not fly for another full 24 hours. Some of our team began processing the latest data on computers installed on the aircraft while parked at the airfield.

Discovered by the Portuguese in the early 1500s, the island was intermittently visited for native food (sea birds, bird eggs, and green turtles, depending on time of year). Domestic animals were released: donkeys, sheep, and goats, some of which thrived.

Ascension was claimed by England in 1815, when a garrison was stationed there in part to ensure that Napoleon Bonaparte did not escape his exile on St. Helena, another volcanic island 800 miles further south. Charles Darwin on HMS *Beagle* visited Ascension in 1836, commenting on the nearly lifeless deserts of the coastal plains. Today Ascension is part of the British Overseas Territory of Saint Helena, Ascension, and Tristan da Cunha—a long name for a few specks of what's left of an empire so vast the sun once never set on it.

Ascension today, as by 1984, had become a kind of signal hub. Already in the 19th century, telegraph cables linking the UK to its colony in South Africa came ashore for signal amplification on the island. It was a strategic spot during WWII—to refuel transport aircraft and as a base from which to fly patrols to look for German U-boats. During the Cold War and today, hydrophones on the surrounding seafloor listen for Soviet and now Russian or other enemy subs. On September 22, 1979, Ascension hydrophones reportedly detected sound waves from a possible nuclear test near the Prince Edward Islands. Such a test was first detected by a Vela satellite, described elsewhere in this book along with details of my involvement in an investigation.

Today the island is one of only four global places from which GPS satellite orbits are monitored and the system tweaked. Another such GPS-tracking antenna in another ocean is on another British island, Diego Garcia. From there, we flew the same P-3 Orion to measure magnetic anomalies over the ocean nine years later. Rockets launched from Cape Canaveral (aka Cape Kennedy) commonly travel south as they rise over the South Atlantic and are tracked from Ascension. Another facility on Ascension—likely very hush hush—involves SIGLINT, short for signal intelligence, i.e. radio-wave eavesdropping. With all those facilities, antennas and radiating cables, Ascension resembles a spider sitting in its web, one composed of cables and invisible electromagnetic links.

Green turtles—a prominent pair on the official Ascension seal—are the largest and most famous denizens. But these 300-400-pound giants only crawl ashore on beaches; the females between May and November. Green turtles are the largest hard-shelled marine turtles. They are unique for turtles by being essentially herbivores, living on algae and seaweed. They live throughout the tropical oceans but are threatened or endangered for various reasons—fishing bycatch, egg collecting for human food, and habitat destruction, among others. When we visited in 1984, they were in decline, but populations measured by the number of beach nests have since rebounded.

We happened to be on Ascension during the right season, and so we had close encounters on at least two occasions. Ascension beaches, where they exist between dark, rocky volcanic headlands, are great for beach parties. Warm, clean water and great sky. On one occasion we shared the beach with several turtles, who went about their egg-laying without paying us any attention. One large female seemed to have trouble heaving herself out of the depression she had made. So several of us helped push the tank-like lady out so she could lumber back down into the surf.

On another occasion, we built a bonfire from packing crate wood (no firewood or driftwood there). Whether from the fire's warmth or light or simple chance, many hatchlings began crawling out of the sand at one place, and many began heading for the bonfire. Of course we picked those up and released them into the dark surf. How many of those have since survived into adulthood, reached after 50 or so years, no one can say. But data suggest that on average, only one in 1,000 makes it.

It so happened that Prince Andrew made a Royal Air Force refueling pit stop in Ascension while we were flying geoscience from the same airfield. The prince was on his way to or from an official visit to the Falkland Islands, recaptured from Argentina just two years prior. His plane had to refuel—too many miles for a non-stop between London and the Falklands, which Argentina has long claimed and calls the Islas Malvinas (Gull Islands). I guess our US Monroe Doctrine didn't apply.

Now, Ascension's 800 or so inhabitants basically live off Ascension coins, stamps, and first-day covers prized by stamp collectors. So I was not that surprised, while researching for this book, when I discovered they had issued stamps and coins to commemorate the princely visit.

At least one person—we know because we met him—was pissed by the royal visit. That was the chief cook at the base cafeteria, a man from Malta, who had gone to great lengths to prepare a special banquet for the prince. However, for reasons

I don't recall now, 40 years later, the prince did not partake. Maybe we peons who fed at that cafeteria consumed some of that food.

We never ran into the prince on his visit. Well, if the truth be told we did once almost run into him. Literally. Several of us in the Naval Research Lab party rented a car on some of our days off. Very few roads on the island, and even fewer vehicles to rent, so we had to settle for a small old ambulance. We were on the narrow road up from Georgetown to Green Mountain. John Brozena was driving—on the left side, as the British do. A black van came down the road, and we and the van nearly either collided or swerved off the slope. We learned later it was Prince Andrew, chauffeured around the island on some official outing, maybe on that day to visit the Ascension Island administrator. Or to gawk at the Green Mountain's cloud forest thickets of bamboo, Norfolk Island pine, and other greenery, nearly all introduced by his predecessor countrymen many decades earlier.

The prince would not have seen many seabirds, who once nested on Ascension in giant rookeries but were wiped out by feral cats when they were introduced in 1815 and were still killing what they could and multiplying on our 1984 visit. However, feral cats were themselves finally extirpated by 2024, and sea birds are recolonizing the island. Offshore, Boatswain's Island continued to host seabird rookeries because cats hate water and won't swim, even for juicy bird chicks or eggs.

How close had we come to creating an international incident, perhaps even a traffic fatality prefiguring that of the prince's sister-in-law Princess Diana in 1997? It wasn't me, or I wouldn't be writing about it now. Who can say how close, but we had to swerve to make room. The road was narrow, and the slope was steep. Maybe Prince Andrew will publish his version—including denials or confessions regarding his later personal life.

I still have a suitcase filled with rock samples I collected on Ascension hikes. I'd be happy to turn them over to any geologist or geology teacher. Had I access to a time machine, I would, of course, fly it back to land on the *Kurchatov* helo deck in 1975 and present them to those Russian geologists.

CHAPTER 12

Mapping Mysteries Beneath the Nordic Seas

Haakon Mosby, who died in 1989 at age 86, was a distinguished professor of physical oceanography at the University of Bergen, Norway. Physical oceanography is about ocean currents, water masses, waves, etc.

In 1980, a research ship built for the university was named after Mosby. In later years I would participate in four cruises on the *Haakon Mosby*, two in 1989, 1990, 1995, and 2010, more than on any other in my career. I was chief or co-chief scientist, employed by the US Naval Research Laboratory on the first three cruises, which explored the ocean floor north of Norway and to the pack ice edge between Spitsbergen and Greenland.

On the last cruise I was invited along as a kind of emeritus—an experienced guest scientist. The *Haakon Mosby* researched an extinct plate boundary in the Norwegian Sea, the Aegir Ridge, named decades earlier by my wife Randi, after the Norse god of the sea. Yes, I had to remind my hosts of that.

At 155 feet in length with a 33-foot beam, the *Haakon Mosby* was smaller than any other deep-ocean research ship in which I journeyed in my career. The vessel later was used for fisheries research and more recently sold to Albania. Smaller generally means more pitching and rolling, and on some stormy days in the Greenland Sea I thought her better suited for the Chesapeake Bay in my backyard. Yes, I sometimes got seasick.

Peter pioneered in the realm of deep discoveries aboard
Haakon Mosby, named for a Norwegian oceanographer.

The 1989–'90 expeditions planned to use side-scan sonar to image the Knipovich Ridge, an oblique segment of the plate boundary between our giant North American tectonic plate and the giant Eurasian plate. Only the Pacific plate is larger in area. By oblique, I mean the ridge was not oriented perpendicular to the direction of plate separation.

Side-scan sonar is much like fish finders and echo sounders on many sports and commercial fishing boats, cargo ships, and passenger liners. However, sports fishing boats such as those used on the shallow Chesapeake Bay use much higher frequency sound that gives higher resolution but is attenuated more rapidly than lower frequencies. That is the same principle under which the thunder you hear is mostly rumbling.

Typical echo soundings are made by measuring the time it takes for a sound ping from the boat to reach the bottom and reflect back to the boat. Water depth is then calculated from the speed of sound in water. The echo sounder was invented after the *Titanic* sank to detect the nearby keels of large icebergs near a ship and was soon applied to measuring water depth.

Side-scan sonar involves measuring not only echo round-trip times but also strength of those echoes. A rough bottom or one covered with hard rocks sends back a stronger echo than a smooth flat muddy one.

Light waves behave the same way. If a flat sea is ruffled by a gust of wind, it turns darker. The sunlight is scattered in many directions by the ripples. Every sailor reads the surrounding water surface. Echo sounding and side-scan sonars both work via transducers, which are crystal-like solid materials converting electric signals from the ship to sound vibrations—the outgoing 'ping.' When the reflected ping arrives, the transducer converts the vibrations back to an electronic signal that is amplified, and displayed on a screen, and recorded by computer for science.

Side-scan sonars involve banks of many transducers mounted around the bottom and sides of a 'towfish' pulled on a cable behind the ship. In this way a fan-like beam of sound is emitted to ensonify—fill with sound—a swath of seafloor.

Work on research vessels can be thought of as 'mowing a lawn.' The ship tracks should be laid out parallel to each other and close enough for the swaths to overlap along their edges. The deeper the water, the wider the swath, but the lower the resolution. A side-scan sonar operating in the shallow Chesapeake Bay maps a narrow swath but easily detects a ghost crab pot and maybe even a beer can, though without being sure it is a beer can. The entire Chesapeake Bay bottom has yet to be mapped by side-scan due to the very narrow swaths, which means high vessel mileage and high cost.

On those early *Haakon Mosby* expeditions, we used the University of Hawaii SeaMARC (Sea Mapping and Remote Characterization) system, and a team came from there to install and operate it. This proved challenging from the git-go because the *Mosby* was smaller than the vessels they had previously worked on. The towfish was very large and hard to launch and retrieve via the stern crane. Plus, the Hawaii team had never dealt with snow or ice coating the decks, a challenge for retrieving and launching their big SeaMARC towfish—the winged platform pulled by a cable behind the boat.

But unlike an image created by echo sounding, a side-scanned seafloor looks different from different angles, so scientists also want additional cross tracks. It's much like comparing LIDAR topographic contour maps of land with photos of that land.

If a five-foot offset seafloor fault scarp, a mini-cliff, happens to run nearly parallel to the side-scan track, you get a clear return. An echo sounder might not see it at all. If the scarp runs perpendicular to your track, your side-scan would likely miss it. Features directly below the survey ship are also missed by side-scan—a circumstance important for our prime discovery on those initial *Haakon Mosby* expeditions. If you drop a small object somewhere on the floor, the best way to find it is to get down and illuminate it from a low angle with a flashlight.

Side-scan sonar sea-floor mapping was new to the Nordic seas in 1989. Up to then, the seafloor was only known from echo sounding, which showed seafloor ridges, peaks, and valleys on the scale of miles across and hundreds of feet of relief. That was enough to 'see' rift valleys and seamounts produced by plate tectonics and larger volcanism: the big picture.

As soon as the SeaMARC seafloor images began to roll out on our paper recorders, we geoscientists began to realize what science had missed. It was a nearly blind person opening his or her eyes upon first getting prescription eyeglasses or opening one's eyes after an eye operation. A large blurry white object became a refrigerator with handles and compartments. A book page had printing that could be read.

Marine geologists aboard the *Mosby* could see individual earthquake faults, probable rubbly cliff slopes, and larger new lava flows along the Knipovich Ridge plate boundary between the Eurasian and North American tectonic plates. Whereas expedition plans focused on such plate tectonic features, the real surprises were, however, not related to plate tectonics. Apparently level or gently sloping sea floors, smoothed by thick sediment covers, now appeared more as a painting, adorned with a rich variety of features with minimal topographic relief and a variety of small scale roughness and variations in sediment type.

POCKMARKS AND METHANE VENTING

The deep-ocean Vestnesa Ridge jutting out just west of Spitsbergen sported along its crest a number of spots, some of which appeared as small pits. We speculated that these might be 'pockmarks' where methane generated by bacteria far below was escaping through the ocean floor. Those bacteria live off buried organic matter, mostly dead plankton snowing down to the ancient seafloor.

Where methane rises to the seafloor, there is generally enough oxygen dissolved in the water to support chemo-synthetic ecosystems, with bacteria living off the methane. Dissolved oxygen is only lacking in a few places, such as in deep

Norwegian fjords, the deep Black Sea, and seasonally in the Chesapeake Bay.

Natural gas, widely used to heat homes and cook food, is almost entirely methane. Methane is also a greenhouse gas that is much more potent than carbon dioxide but does not last too long in the atmosphere. Moreover, except in very shallow waters like Chesapeake Bay, any methane escaping into the water column is oxidized on its way up, before it can escape into the air.

In 1998, almost a decade later, I had the chance to see with my own eyes one of those putative pockmarks we had discovered on *Mosby*. That was my second dive on the Russian Mir submersible. Yes, here was such a pockmark supporting a chemosynthetic ecosystem, with white mats of bacteria, starfish, eelpout fish, and other organisms. (The Mir dive expedition is reported in another chapter.)

Science, wine and international bonhomie aboard the *Haakon Mosby*.

ICEBERG MARKINGS

There was enough ice-free water north of Vestnesa Ridge for us to make another and totally different discovery: iceberg plowmarks. Thick icebergs or floating ice shelves have deep keels and scrape shallow troughs into sediment-covered seafloors. The phenomenon depends on the water depths at that time and the thickness of the icebergs or shelves.

Our side-scan images revealed two sets of plowmarks, an older deeper set and a younger, shallower set, one that looked to have erased part of the older set. Both sets of grooves were straight, as if made by a giant plow or comb. Isolated icebergs tend to leave curved scratches. So what we saw must have been scrape marks made by a floating glacier-generated ice shelf or wide tongue, or many icebergs packed together. There are few icebergs in the Arctic Ocean, and nowhere are they thick enough for their keels to reach the ocean floor.

What we found was evidence from past ice ages when floating ice shelves or tightly packed armadas of bergs covered parts of the Arctic Ocean—much like ice shelves flowing into the ocean from Antarctica today.

Additional seafloor evidence for past floating ice sheets has been found in the 30-plus years since our discovery. The older, deeper set of our plowmarks was likely formed during the penultimate glaciation (aka ice age) of North America and Eurasia, about 140,000 years ago. That's before Homo sapiens left Africa. Some scientists think the Arctic Ocean was then completely covered by a floating ice sheet, leaving that ocean totally dark.

By sheer coincidence, the buried valley carved by the Susquehanna River during that ice age was discovered by others under the Chesapeake Bay during the same years (1989–'90). More about the Chesapeake in a later chapter.

SOUPY MUDFLOWS FROM LAST ICE AGE

The continental slope between Spitsbergen and northern Norway looked smooth from echo sounding data. Our SeaMARC side-scan images revealed that the slope was actually decorated by numerous long tongues of sediment, too thin to show up as relief on echo sounding data. Their existence surprised and excited all the experts.

These are now interpreted as soupy 'glacigenic' mudflows starting from the edge of the former Barents Sea ice sheet, which carried and pushed masses of sediment to its mouth near the continental shelf edge. Moreover, subglacial streams did some of this work. This Barents ice sheet was grounded below sea level like the modern West Antarctic ice sheet and was the first to collapse as the last ice age ended. A possible tipping point model for what might happen in West Antarctica under modern greenhouse warming and climate change?

SOME SPECIAL MUD

By far the most memorable and most subsequently investigated discovery we made on the 1989–'90 SeaMARC side-scan expeditions on *Haakon Mosby* was a sea-floor mud volcano. Such features were not previously known from that region and this one is practically the only one there.

Scientists get their hands dirty with the
feel of volcano mud from down deep.

Our discovery illustrates that simple good luck can play an
important role in science. It happened at the start of our second
season, 1990. We were headed from Tromso to our research
target, the Knipovich Ridge plate boundary, and decided to
launch our big SeaMARC towfish enroute to test the system.
When I say we, I mean the University of Hawaii SeaMARC
team headed by Alexander "Sandy" Shor.

On the way, we saw a rounded feature about two-thirds of
a mile in diameter centered below our track. Now, side-scan
sonar lives up to its name and does not reveal much if
your transducers are directly above the feature. There was
nothing evident on the echo sounding; nothing rising from
the seafloor. Features like that are commonly computer-
generated artifacts, and one of my postdoc fellowship
guys onboard, a seismologist who knew much more than

me, decided it was an artifact. I was not totally convinced, so we looked very carefully at our echo soundings across the feature. To my happy surprise, the record suggested that the feature actually rose several fathoms above nearby seafloor—enough to rule out a computer artifact.

We didn't have time to turn around and run a line near but not over the feature to get a good side-scan image. The discovery remained for five years until, in 1995, we found the funding to return to take sediment samples and measure heat flow from the same vessel. The box core recovers not just a cylinder of mud but more than a square foot area and a one-foot-thick sample of what was down there. Cold gray mud, but with thin, curly spaghetti-like things growing up out of the mud. These were tubeworms, part of this chemosynthetic oasis. We jokingly named named the tubeworms Erica's Hair after after a young female scientist on board.

The mud volcano we discovered and in 1995 verified was named the Haakon Mosby mud volcano. Google searches, especially Google Scholar, will likely prove this is the most-studied of numerous seafloor volcanoes under the world's oceans. And those tubeworms turned out to be a new species; you guessed it, they were named after Haakon Mosby, whose name adorns not just a ship but also a mud volcano and a new worm species.

Later cruises by French and German research ships showed the highest peaks of this mud volcano tower only 30 to 40 feet above the surrounding seafloor. The feature does not have the shape people associate with volcanoes; you can't build a miniature Fuji from soupy mud. This mud volcano resembles a 3,000-foot in diameter cow pie!

In 2000, the University of Bergen awarded me an honorary doctorate for my research, especially in the Nordic seas. Did the Haakon Mosby mud volcano play a major role in this honor? I am sure of that. I had Lady Luck on my side, but I did have the good sense to find evidence that our sonar image was not a computer artifact.

On our 1995 *Mosby* expedition, we found more than tubeworms in that gray cold mud. After one of the cores arrived on deck, I grabbed a handful for a closer look. There was something white moving across the mud surface. I thought that maybe it was another organism. But the pale object was shrinking and bubbling. It must be methane hydrate, I concluded. I had only read about it, but never had I seen any up close.

Methane hydrate is an icelike material composed of water and methane. It is stable under high pressures but not under an atmosphere except very far below zero. Large volumes of it underlie sediment piles along many continental margins, including off the southeastern United States. It is a possible source of natural gas, but under climate and ocean warming also may be a future contributor to greenhouse warming if the hydrate separates and the methane gets into the air.

Supposedly, you can hold a chunk of the stuff in the palm of your hand and light it, burning off the methane and leaving some water in your hand. I tried that with gas hydrate from the same mud volcano on the 1996 *Logachev* cruise, but I didn't succeed. Maybe the Russian matches were bad.

While my stomach would disagree, I cannot speak too highly of my overall experience on *Haakon Mosby*. The vessel was operated by only four officers and four crew, all

Norwegians. They all spoke English, but they accepted me as Norwegian-speaking. I had learned their language a decade or so earlier as a visiting scientist at the University of Oslo. Except during station work or approaching or departing harbors, I hung around on the bridge. At least on the later expeditions, the chief engineer monitored the invariably hot engine room instrument panel from the comfort of his cabin. Probably routine on many modern ships.

The 1995 investigation of the Haakon Mosby mud volcano was the science highlight of that expedition. The other highlight had nothing to do with research. On the way back south to Bergen, we first secured all our science equipment, washed mud off the deck and tidied up the lab. Then we stopped briefly in Tromso in northern Norway to pick up three fine ladies, none of whom had ever ridden on a research ship. Two of the ladies were wives of shipboard scientists: my wife Randi and my sedimentologist Fred Bowles' wife Sandy. The third, Soula, was the Greek-American fiancée of research computer man Chris Jones.

The three would ride along south through the famously scenic inside passage back to Bergen. There is one area where ships have to be out on the open North Sea. On some previous transits back to Bergen, I suffered for hours in the most extreme pitching of my ocean career. The shallow water depths close to the Norwegian coast cause storm seas to pile up even higher. However *mirabile dictu* on this passage. Even the open Norwegian Sea was practically as calm as a mill pond. My wife began to doubt my tales of stormy seas.

The inside passage has many twists and turns. The skipper knew them all by heart.

At some spots the cliffs seemed close enough to touch. We stopped close to the fjord-edge summer cottage of my Norwegian colleague and friend Prof. Eirik Sundvor. He rowed ashore and back to the ship. I guess the university was okay with that stop, and another one elsewhere where some booze cases were delivered to our vessel by an older man in his rowboat. A casual scheme to bypass the astronomical duties on imported spirits, unthinkable in the United States.

At the time wine and beer were okay in modest quantities on Norwegian research ships. Later, they copied the Puritanical American zero-booze shipboard policies, sometimes violated if you kept your mouth shut.

Sundvor had initiated a tradition of chief scientists (in 1995 that was Erik, Kathy Crane, and me): treating the ship to wine every Sunday. On this exceptional passage we dined and toasted on the aft deck while feasting on the spectacular scenery sculpted by recent Scandinavian ice sheets.

I have two profound 1989 memories from the northernmost point—near the pack-ice edge in the Fram Strait between Greenland and Spitsbergen—and the southernmost point, Bergen.

Geophysical cruises to the far north are planned for late August to mid-September, when the pack ice edge has retreated to its yearly limits. But locally, this depends on winds blowing sea ice south or cramming ice floes together farther north. Of course, the oceans get stormier and daylight shortens as summer grades into fall.

After flying to Oslo and then a short hop to Bergen to join the expeditions, I was generally greeted by cloudy skies and wet streets. Bergen is much farther north than Seattle but resembles it in terms of climate due to a branch of the Gulf Stream. I usually stayed at the small Park Hotel, a close, rainy walk to the Bergen University Geoscience building, where I often enjoyed reasonably priced open-faced sandwiches and watched TV in the student cafeteria. On some trips to Bergen I was put up by Eirik and his wife in their nice suburban home, and I would ride to the university with him.

In 1989, we experienced the first SeaMARC expedition and departure delayed and delayed again due to technical problems that I have since forgotten. Was it parts expected but shipment delayed? I spent hours in the student cafeteria watching TV reports about events in Eastern Europe, especially Germans trying to escape the GDR and actually protesting in some cities like Halle. Would Gorbachev intervene as the Soviets did decades earlier in Hungary and Czechoslovakia?

Finally, we were underway on *Mosby*. The SeaMARC side-scan surveying began. Radio news, mostly marine weather forecasts, included world news, even events behind the Iron Curtain.

One night I recall vividly, we were in the Fram Strait, the deep passage between the Atlantic and Arctic oceans. It was late October, pitch dark and well below zero. A few snowflakes swirled around the lights. Due to sea smoke, the *Mosby* floodlights did little to illuminate the sea ahead for floating ice. We depended entirely on radar, which showed the edge of an ice pack just a few miles ahead. But Captain Faeroy kept the nonchalant cool look common to

Norwegians. I was excited but not worried, even though the skipper had probably never been up here this late before.

The captain not only studied the radar but tracked ocean temperature, and we were still in the now-cooled but saline Gulf Stream water. This is where that water is finally cold enough to sink, due to higher salinity below the fresher colder Arctic surface water. I thought about the water mass we were floating in having once been the Gulf Stream that passed northward east of Maryland. Some then had sunk off Greenland, eventually to flow south well offshore Maryland—but deep down as the Western Boundary Undercurrent. Our grand ocean plumbing system of today, but maybe not of the future.

Finally docked back in Bergen, I checked in at the familiar Park Hotel and flopped down on the bed, happy to be back on solid ground. Happy the expedition was successful. Happiest of all about soon to be back home on the western shore of the Chesapeake. I flipped on the Norwegian TV. A reporter seemed to be interviewing two East Berlin schoolgirls who had casually walked through a nearby gap in the wall into West Berlin, without telling their parents. The reporter, who likely talked to the girls in German, was relaying that to us. My Norwegian was very good, but not perfect. I must have missed something big.

So I turned on English language TV or maybe radio. What I saw and heard on Norsk Fjernsyn television was true.

History was made on that day, November 9, 1989: the fall of the Berlin Wall.

CHAPTER 13

In Search of Tectonic Plates (and Coconuts)

Diego Garcia is an atoll in the Chagos Archipelago in the western Indian Ocean. This 12-square-mile tropical, reef-rimmed island was discovered by the Portuguese in 1512 but not colonized until 1793, first as a French leper colony and later, with slave labor, a coconut plantation.

A view at Diego Garcia, the largest island of the Chagos Archipelago, remnants of the British Empire.

The British—who had tried and failed to establish a colony in 1786—grabbed it during the Napoleonic wars, but the local coconut and copra plantation economy continued. Many of the workers had been imported from other Chagos islands and continued to speak French, even developing their own Creole dialect. Formally, it has long been a British Indian Territory. After the large Indian Ocean Island Mauritius—also 'acquired' from France by Britain—gained its independence, Mauritius laid claim to Diego Garcia, a legal dispute resolved only in 2025 when the UK recognized Mauritius' claim to the entire Chagos Archipelago. The UK and US will continue to lease the joint military base on Diego Garcia for more than $100 million a year. (I had visited Mauritius as a grad student watch stander aboard RV *Chain* in 1964 and for a Project Magnet refueling stop in 1968.)

The history of Diego Garcia changed dramatically during the period 1968–1973, when the US collaborated with the UK to build the bomber and ship-supply base on the atoll. For that purpose, the native Chagosian, Creole-speaking plantation workers and their families had to be expelled. This forced expulsion and resettlement of about 1,000 people (in Seychelles, Mauritius, and Chagos islands) took place in 1971. In 2019, the International Court of Justice declared the expulsion illegal, but the UK claims the Court has no jurisdiction. Legal contests, including compensation claims, are ongoing.

Today, this militarized atoll has a population of about 4.200, comprising military personnel and contractors. No, the public isn't welcome there to enjoy the coconut groves, white beaches, and land crabs, along with superb fishing and diving.

My visit to Diego Garcia in 1993 was connected to Charles "Chuck" DeMets, who had just completed his postdoc with me at the Naval Research Laboratory. (He later became a tenured professor at the University of Wisconsin.) Back then, DeMets was still in contact with his Northwestern University thesis advisor, Richard Gordon, a prominent geophysicist with a keen interest in plate tectonics in the Indian Ocean area.

DeMets measured present and recent motions among and between all tectonic plates. When global plate tectonics was first proposed (1967–'68) geologists considered Australia and India to be on the same Indian Plate. The northern edge of this plate is marked by the Himalayas, where the Indian continent, part of that plate, is colliding with the Eurasia plate.

With later research, based on rates of plate motion, earthquake seismicity, and other factors, it began to look like the Indian Plate was actually starting to break in two. But the boundary between a new, smaller Indian plate and an Australian plate was fuzzy—marked just by widely scattered earthquake epicenters. If there were two plates, the Indian-Australian plate boundary would have to intersect the Mid-Oceanic ridge plate boundary, which runs south towards Antarctica from an exit near the Gulf of Aden. Meanwhile, it was recognized that the Africa Plate is also two plates—today called the Nubian plate in the west and the Somali plate in the east. (The boundary between these two follows the East African Rift, the likely cradle of human evolution.)

So where was the triple plate junction? Maybe an area versus a point, where the Somali, Indian, and Australian plates joined? Linear magnetic anomalies could tell, and the easiest and quickest way to measure those was by air. The Naval Research Lab had a P-3 Orion with a magnetometer, and Diego Garcia, with its long runway and other support, was located close to the area of interest so we would not burn up a lot of fuel in transit. The Lab could not justify funding the entire project, but for policy reasons the National Science Foundation could.

Gordon and DeMets wrote the proposal, and the anonymous expert reviewers gave it high marks. However, the National Science Foundation for PR reasons doesn't send funds to another government agency. So I had no hope of them paying for my participation on the flights nor back at the office analyzing data. However, they did agree to pay for aircrew travel expenses. There was simply no such aeromagnetic capability in academia.

Getting our P-3 Orion from Maryland to distant Diego Garcia, and back, was no trivial matter. It's a long way, so fueling stops were required. The shortest way is eastward, but that would mean crossing various African nations, some hostile to the US. Another issue with Diego Garcia base facilities— which were superb, but basic science research was lower in priority than military objectives. Lucky for our project, there was geopolitical calm. The first US Iraq war against Saddam Hussein to vacate Kuwait started not long afterward—with bombers using the runways.

With the lack of motels and restaurants, we civilian scientists stayed in the Bachelor Officer Quarters with the officers. The pilots were not allowed to fly every day, so we had days on the ground. Chuck and his ex–prof used some of this time to process the magnetic and navigation data in the Bachelor Officer Quarters or in the parked P-3. We frequently discussed this and other tectonic plate problems. On some free days we three rode rental bikes around to the far side of the atoll to view the remains of a French coconut plantation house.

I hiked on the wide beaches and collected abundant tropical shells; some species were not widely available in museum collections. I later donated my shell collection to the Calvert Marine Museum in Solomons, Maryland, where paleontologists could examine shells of living species to compare details with related extinct ones. A volunteer who since died identified and labeled and cataloged the donated shells. Thus, a bit of the tropical Indian Ocean resides in a southern Maryland museum. At least today, the relatively pristine native marine and terrestrial wildlife on and offshore Diego Garcia is strictly protected.

Our quarters were close to the beach, and coconut palms grew nearby. I noticed plenty of coconuts lying around on and near the beach. There was no evidence anyone had tried to crack any or drill holes into one of the eyes to drink coconut milk. Throughout the tropical world, street vendors sell coconuts on the street. As a fan of foraging edible wild plant foods (after reading *Stalking the Wild Asparagus* and other works of Ewell Gibbons, I couldn't resist the temptation to try to break open some of those coconuts). It wasn't easy but with the help of concrete rubble I started to crack open and consume some coconut meat and juice. Chuck happened by and decided to learn from his past postdoc advisor. He improved on my method, commenting that students were expected to someday surpass their masters. We were harvesting coconuts as those Chagosian plantation workers had done many years before— except they would have scorned our primitive methods.

Flying back and forth at only 1,000 feet above the ocean, our flights collected valuable magnetic data. The air was nearly calm; no storminess in the Intertropical Convergence Zone. The lack of decent winds near the equator was a bane to old-time sailing ships, but not to us.

Upon project completion, the trip back to the States was again a long way. I first flew military transport from Diego Garcia to Subic Bay Naval Air Station in the Philippines. Subic was later destroyed by the eruption of Pinatubo and the land returned to the Philippines. Then a commercial flight to Guam, and from there via Anchorage (yes, Alaska) to San Francisco and back home. I could have stopped at hotels along the way, but I was anxious to get back. My total travel time was 51 hours, including waiting to change flights at airports. I have trouble sleeping on airplanes, but I slept soundly my second night in the air.

The magnetic data we collected were analyzed and published in the prestigious *Journal of Geophysical Research*. DeMets or Gordon were first authors. As more of a facilitating scientist,

I was only third author. These papers are still cited today and I take pride in my participation. This project was a strange, 30-years-later-déjà vu in my career. My very first research publication (Bowin and Vogt, 1966, based on data from the 1964 RV *Chain* expedition discussed elsewhere in this book) had me co-author of a paper dealing with linear magnetic anomalies in the same journal and also about magnetic anomalies in the western Indian Ocean.

Of course, that was before global plate tectonics was discovered.

CHAPTER 14

From Ocean Commotion Into the Still Depths

The world's largest known sub-marine landslide occurred roughly 8,150 years ago, 62 miles off the coast of Norway. A 180-mile-long length of the continental margin caved away, moving a staggering 840 cubic miles of the coarser sediment up to 250 miles and finer particles 500 miles out into the Norway Basin.

To put this volume in perspective, 840 cubic miles is roughly 25 million times larger than a modern large landslide along the Chesapeake Bay's Calvert Cliffs. There is some recent evidence for an earlier mega slide at Storegga during the last Ice Age 20,000 years ago. The Norwegian term Storegga can be translated as 'big edge.' The fishing community has long known of this abruptly steep drop-off along the continental margin.

The Storegga Slide created a megatsunami that washed inland and left behind tell-tale sand deposits. Since the 1980s, these deposits have been identified and age-dated on the coasts of Norway, Scotland, the Shetland Islands, and the Netherlands. The waves raced up to 50 miles inland and 13 feet above present sea level, inundating and destroying Middle Stone Age settlements. The slide was likely triggered by an earthquake, dislodging the masses of loose glacial deposits left by repeated ice sheets.

The large offshore Norwegian oil industry was, and still is, very interested in the Storegga Slide. Could it happen again, perhaps triggered by seismic prospecting or by drilling? The consensus seems to be that given that failed sediments were dumped there by past Scandinavian ice sheets, there is little risk until after a future ice age. However, new evidence for an earlier Storegga Slide changes the risk estimates, which were based on a one-off scenario.

With Naval Research Lab and academic colleagues, I studied the various kinds of Storegga landslide deposits—geologically very young and thus fresh—blanketing the Norway Basin ocean floor. The relevant data consisted of side-scan sonar images and bottom topography, both collected from several different research vessels, especially the RV *Maurice Ewing* of the Lamont-Doherty Geological Observatory, chartered by the Office of Naval Research to do that seafloor mapping.

One day in early 1999, Hank Fleming, my Naval Research Lab branch head, asked what I would do if I had a chance to do some marine geological research from Submarine NR-1, a Navy nuclear-powered, ocean engineering and research vessel. I was surprised and said I would drop everything and give it some thought. Hank needed a proposal and fast.

I had previously read about this small, unique sub, the brainchild of the late Admiral Hyman Rickover, known as the father of the nuclear Navy. This vessel could dive deeper (at least to 3,000 feet) than the much larger conventional US attack subs such as the Los Angeles class boats and the even larger strategic 'boomers,' which could fire ballistic missiles and stay submerged for months. NR-1 could collect things from the seafloor, deploying a so-called object-recovery claw, and deposit them in a basket. On a level, relatively firm bottom, the sub could even drive around on wheels. NR-1 had many still cameras and TV monitors, and several portholes from which to inspect the ocean bottom. There was also a side-scan sonar observing the seafloor on both sides.

The NR–1 was the smallest nuclear sub in the American fleet, deployed for research, recovery, and surveys.

NR-1 was 400 tons and 148 feet long, with a beam of just 12.5 feet. Much of its length was devoted to the reactor and turbine generator. Its speed underwater was only 4.5 knots and 5.2 knots at the surface. The sub was built at the Naval Submarine Base New London by the General Dynamics Electric Boat Co. After its January 1969 launch, NR-1 had had a 30-year-long distinguished career before 1999. Many of the deployments were classified missions, and our cruise was obviously not. Among popularized missions were discovery and collection of seafloor debris from the 1986 Space Shuttle *Challenger* explosion and Bob Ballard's 1995 exploration of the wreck of HMS *Britannic*, *Titanic*'s sister ship.

NR-1 carried up to 13, including three officers and two scientists or other specialists and assistants. Everyone who crewed on NR-1 was nuclear trained and personally screened by the director of the Naval Nuclear Propulsion Program. My limited experience on this unique sub only underscores its motto: The World's Finest Submersible.

My thoughts raced back to the Storegga Slide. NR-1 could not dive to the depths of the Norway Basin to look at the landslide debris I had been studying. But it could drive along just above the slide headwall—the underwater cliff slope from which those 840 cubic miles of sediment had broken loose. The Norwegians were, of course, very interested and still concerned about possible new slides. So we could look for any cracks or recent movement.

There was, however, a problem: Norway was totally nonnuclear and did not want nuclear-powered subs or armedvessels docking in their ports. There were negotiations that led to their permission for NR-1 to dock in Bergen, so they made an exception provided that a Norwegian geologist would be aboard. Given that there is barely room for one scientist plus one assistant, it was decided that the Norwegian (actually an Icelander named Haflidi Haflidason) and I would each have one week under the sea. Each of us would have an assistant, a midshipman from the Naval Academy in Annapolis.

After my undersea week, I would exchange places with Haflidi in an open-ocean transfer on a Zodiac between NR-1 and its tender (mothership). Then I would hang around on the MV *Cory Chouest* until his week was up and we all would return to Bergen as NR-1 was towed back.

Hank Fleming, my boss, and I flew from DC to New London, the US Sub base, to inspect NR-1. With little of the sub beyond the conning tower exposed as NR-1 lay alongside the dock, she looked pretty small. We were shown around the sub, except for the reactor area, which is only entered by the crew for inspections. A highpoint of the tour was the long freezer, from which we would—once underway—pick and choose from prepared meals for heating in the microwave. And there we spied the notoriously popular and thus short-lived Klondike ice cream bars.

There was no real galley, just one cubicle where four could sit around a table. I would have to 'hot bunk' in a narrow, cot-like space sandwiched among various pipes and valves just forward of the entry to the reactor. While underway, I recall being worried I might accidentally move one of the valves while climbing in or out. NR-1 had no shower, so washing oneself was with wet sponges. The toilet seemed technically complex, so I had to relearn the procedure once we were underway from Bergen. It reminded me a bit of *2001: A Space Odyssey*, in which a solitary space passenger confronts a sign reading: Instructions For Use of Zero Gravity Toilet.

I already knew, or was reminded, that Bruce Heezen, a famed ocean-floor geologist, died of a heart attack in NR-1 while researching the Reykjanes Ridge plate boundary in 1977. They preserved him in that same freezer so the mission could be completed. He was of the first of the post-war generation of ocean geologists who, due to their small number, were rather braggadocious. I recall no matter what seafloor geology or geographic feature I asked him about, he had discovered it. Of course he was a pioneer and giant in ocean geology.

The sub would be towed empty across the Atlantic from New London to Bergen by the mothership. I recall the towing hawser being several inches thick. I would next board NR-1 in Bergen, Norway, and be towed by *Corey Chouest* about 100 miles above the Storegga Slide headwall area.

NR-1 was turned loose, and we soon descended from ocean commotion into the still depths. We came within sight of the seafloor around a half mile down. The large depth gauge—similar to a clock—showed the depth. The dial went up to 6,000 feet, but the nominal depths we were not supposed to exceed was at the 6 o'clock position. I suppose NR-1 has operated below 3,000 feet but I didn't ask, figuring the answer was classified.

The Storegga headwall slope sported widely scattered rocks and white sponges. No cracks or evidence of escaping gas or water were seen. We glided across a large patch of deep water corals, a ghostly thicket about 5 to 10 feet high. There was a small level plain on which we descended and drove around on wheels. Driving back across the track provided information on bottom sediments to contrast with my Mir dives in the same general area the year before. On NR-1, one can't get up close and personal with the seafloor. Even when the sub was close to the bottom, our portholes were too high above the bottom to make out details. Visibility even in clear water is limited, and the abundant marine snow doesn't help. Imagine an airliner pilot on a runway reading something on the tarmac in snow flurries. And air is clearer than water.

There were a few glitches. Those deep-water corals should have been sampled, but the object-recovery claw system was, alas, not operative. A few days into our dive, the vessel's commander, Charles Richard, came to me and said: "Dr. Vogt, we have a small problem. Our carbon monoxide scrubbing system is down. This means CO is building up in the sub, but won't reach dangerous levels before we surface to transfer you to *Cory Chouest*. It's up to you if you want to continue or abort this dive at this point."

I said I knew about the CO_2 scrubber but not that carbon monoxide would build up inside a nuclear submarine. He explained that we have 13 flatulating males on board, each emitting intestinal gas. It was embarrassing to have learned more about human biology from a nuclear submarine officer. I decided that we should continue with the mission, and so we did.

People ask me if we saw any fish. From portholes I saw none. However, on video, there were usually several two foot-long fish, probably cod, visible due to flood light illumination, swimming under the hull. Were they always the same ones? We could not tell. Were these cod there for protection or some other reason? Only the fish knew. Of course, this mission was about geoscience. There were no ichthyologists onboard.

It was cool to contemplate that aside from food, we were powered not by ancient stored solar energy from fossil fuel but from radionuclides with long half-lives-produced by an exploding star before the solar system was born. More down to earth, much that goes on inside a small sub has to be somewhat sustainable. For example, space is very limited, so one crew member was tasked to compress our paper waste— wrappers of TV dinners, and Klondike ice cream bars, for instance. He put such trash into a press (compactor) every day and squashed it into a small block.

When my underwater week on NR-1 was up, it was time to grab my onboard laptop (an IBM Powerbook; laptops were not yet thin) and a waterproof bag with a few key belongings, climb up two fixed ladders, and wait inside the closet-sized space of the conning tower. I had, of course, donned a PFD. A crew member standing just outside on the small landing area on the hull would open the hatch, let me out, and with no delay I would get into the small Zodiac pontoon boat.

The Zodiac had reached NR-1 from the *Cory Chouest* and was there already, heaving up and down and back and forth alongside the sub. Lucky for me, the notoriously rough Norwegian Sea was relatively calm that day. I said relatively. The key was to time one's jump to the crest of a swell and the Zodiac's proximity to the hull. Although not famously coordinated, I managed that, *mirabile dictu*, and we were off putt-putting on the open ocean. The open ocean up close and personal. Maybe a half-hour to *Cory Chouest*, having done that only rarely in my long career. Haflidi had to board NR-1 in the same manner and get back to the mothership the way I did.

At the end of the mission, we were both onboard comparing our Storegga Slide seafloor observations as NR-1 was towed back to Bergen. By then the approach through a calm fjord and around forested islands was very familiar to me after three research cruises (1989, 1990, 1995) on *Haakon Mosby* (University of Bergen) as told earlier in this book. On our approach to the inner harbor, a massive concrete U-boat pen—a reminder of WWII German occupation—was in plain view. As we pulled along the dock, I wondered if it was the same one where our *Atka* icebreaker had tied up 33 years earlier—and where the Norwegian-American girl named Randi that I had been dating and would later marry was standing and waiting for my arrival.

Back at the Naval Research Lab, I had to find time to pore through a box of videos, side-scan sonar CDs, and notebook observations. The audio recordings by the officers on watch were included with what they observed on the seafloor below while I was asleep or eating lunch. Yes, I had briefed them on what to watch out for. I had just too many other duties, so I offered this data trove to a postdoctoral fellow. I had 14 of them through the years, from one to several men or women at a time, each for one or two years. Some we later hired, depending on our need, funding, and their qualifications. Something like dating before marriage. In the case of the NR-1 data, I turned the material over to geologist Brian Parsons, who analyzed everything and, with my and Haflidi's contributions, published the results a few years later.

One day soon after Brian began to wade through the data, he rushed into my office excited and grinning. What could be funny about the NR-1 data? He asked if I had listened to any of the audio. No I had not. I hadn't even realized all that was recorded. The audio recordings revealed that when I was up there, the discussions involved nuclear reactors, geophysics, and such. All technology and science.

In my absence there was plenty of sailor talk, "f___this" and "f___ that." Obviously, these officers did not suspect I knew sailor talk and would never be offended hearing it from them. Of geological interest were larger rocks on the seafloor, first brought from Norway by the ice sheets and carried down in the slide. What were those BFRs in the officers' audio and notes? Why obviously, "big f_____" rocks. In our technical paper you won't find any mention of BFRs. I offered the NR-1 audio tapes to a neighbor working for Smithsonian History and Technology. The tapes tell an amusing vignette of US Navy culture at sea. Alas, the tapes were likely scrapped after I retired from the Naval Research Lab.

PART IV

From Cruising With Soviets to Cruising with Russians

Of all my voyages to distant places and ocean depths, the most farfetched were defined not by our destinations, exotic as those were, but by our fellow voyagers. In 1975, the hot heart of the Cold War, when the Soviet Union and the United States were rivals for world dominance, I was invited to join our 'enemies' on a scientific expedition.

A brief thaw led to USSR and US scientists becoming guests on each other's nonmilitary research ships. On the Soviet ship *Kurchatov* we three Western scientists collaborated with our USSR colleagues. I found some high-level Russian bureaucrats not so cooperative. No one did.

On the Soviet ship *Kurchatov*, we worked and studied together through the South Atlantic Ocean to the most remote inhabited island on the Earth—though our interest was not land but the ocean floor beneath it.

Eighteen years later, the disintegration of the Soviet Union seemed to promise a future of international cooperation. Yet in the new era of openness, I found Russia a less cooperative colleague.

I planned to charter a Russian research ship and help fund their scientists who, with our US team, would investigate the Kara Sea, where the USSR had dumped lots of nuclear waste. Our chartered Russian research ship never left the dock, scuttled by Russian authorities.

Change came fast to Russia in those tumultuous days. By 1998, the year of my third expedition, the Russians were renting out their Mir submersibles to whoever might be interested in exploring the ocean bottom. My turn came shortly after the two submersibles enabled the filming the of the blockbuster American film *Titanic*.

Our interest in my two 10-hour dives focused on present sea-floor processes and past geological events. Seeing close up and personal such anomalies as the Harken Mosby mud volcano and small seafloor craters called pockmarks where methane bubbles out.

I had been researching the seafloor and below from the surface off and on for the last 34 years. Now I was seeing the bottom for the first time.

CHAPTER 1

Avoiding Politics Aboard the Kurchatov

One office day in early 1975, I got a most unusual call. It was from Alex Malahoff, a fellow geoscientist I knew at the Office of Naval Research (ONR). The office funds basic university research of potential future Navy interests and was created soon after WWII to support science important to US national interest. Many talented scientists employed during the war would otherwise have sought employment elsewhere, and their talent would be lost to our national interest. ONR preceded and inspired the National Science Foundation as a means of funding valuable basic research not feasibly conducted in the commercial sector.

But why would Malahoff be calling me at the Naval Research Lab, where research is done by civil servants, not college professors?

He explained that as a result of a small current thaw in the Cold War, US and Soviet ocean scientists were invited to participate as guests on adversary research cruises. Three US academics had been lined up to ride on the Soviet ship *Akademik Kurchatov* on an expedition to the Equatorial Atlantic from late April to early June. But for some reason, one of the academics had reneged, so Alex had to find a replacement on short notice. He appreciated my science qualifications and hoped the Soviets would accept an employee of a US military service. I had also taken two years of scientific Russian at Caltech 15 years earlier.

We were aboard the Soviet vessel *Kurchatov* in May, 1975, on the 30th anniversary of the German surrender and the end of World War II—with Russians toasting their Great Patriotic War victory.

Would I agree? Would the National Research Lab let me go? And what would my wife Randi think, given that I was already committed to two months as co-chief scientist on Leg 43 on the drill ship *Glomar Challenger* in July and August?

I was no fan of the Soviet Union and communism and had never set foot in the USSR. Yet here was a rare chance to experience how the other side conducted ocean science on a non-military vessel. What food did they serve on such ships? Would anyone try to sell me on their type of government?

After talking it over with Randi, I called Malahoff the next day and agreed to solve his problem. And 1975 became the year I spent the most married time away at sea. Our two boys were three (Jason) and seven (Erik). I wondered if that reneging academic was married to a less tolerant wife.

As for the Soviet exchange scientists on our academic research ships, their equivalent of Alex Malahoff had other worries: Could he or she be sure the chosen scientist would not defect? That happened once, in Hawaii.

I think Malahoff shared the names of the two other Western exchange ocean geoscientists: John Malpas, a Canadian petrologist (igneous rocks) from Newfoundland; and Susan Humphries, a British-born, Woods Hole researcher of sea-floor hydrothermal systems (hot water springs).

In late April I was on a flight to Madrid, from where I would fly to Las Palmas in the Canaries to board the *Kurchatov*. Las Palmas was used by numerous Soviet fishing and fishing-factory ships active in the far eastern North Atlantic. So was Dakar, Senegal, where I would later disembark.

I arrived in Las Palmas during the Easter holidays, when the islands are packed with sun-worshiping Northern European tourists. I think my office had failed to book me into a hotel. I can't recall if I found a hotel room or if I could already get on board. My experience on our vessels is that cooks and stewards want time off in port, so they are not fond of scientists and grad students using the ship as a cheap lodging.

I wasn't assigned to any labs or watches, so I explored the ship and helped on science watches when possible. I interviewed the computer man and got samples of paper seismic records.

The equator-crossing events were only somewhat similar to those on *Chain* 43 a decade earlier. I had forgotten my card certifying me as Shellback. They saved us three foreigners to the end, when the pool we got tossed into was good and dirty. The captain had ordered Neptune and his henchmen not to be too rough on us; apparently bones had been broken at such hazings. It did not help that the henchmen were somewhat liquored up by the time they got to us. But it was all good fun, we showered and we enjoyed some booze at the party that followed.

I spent hours on the highest deck getting a tan and studying Russian. I planned to give a shipboard science seminar and wrote out drafts for my Russian peers to correct. My seminar lecture was on hotspots like Hawaii and Iceland, with postulated mantle convection plumes perhaps causing the three volcanic islands, two of which we visited.

However, all Soviet geology was then dominated by Vladimir Belousov, who rejected plate tectonics. By 1975, Western geoscience had accepted it. Holdouts were those studying continental interiors where the land only went up or down. American petroleum geologists for years agreed with Belousov for similar reasons. But not likely still in 1975.

Soviet geoscientists could cite western publications on plate tectonics but not publish their own evidence. So here I was talking about hotspots to scientists who had not even accepted tectonic plates. Yet my seminar went well. After my presentation, however, marine geologist Gleb Udintsev teased me about thinking I had found the Philosopher's Stone.

We three Westerners tried not to provoke political discussions. We knew our peers had some Communist Party minders on board. They had occasional closed meetings to which we were not invited. John Malpas, the Canadian geologist, went so far one day in the galley to bring up Czechoslovakia, whose 1968 revolt the USSR had crushed. I don't think he got any bites. I once tried to provoke a discussion by observing that our system had problems, such as lack of medical care if you were poor. I hoped they would then acknowledge problems in their system. However our Soviet colleagues only admitted some of their roads were bad.

As on our ships, movies were shown. Some were black and white Stalin-era films of peasants singing while harvesting grain. Our colleagues seemed a bit embarrassed by such films.

As it happened, *Kurchatov* was at sea on May 8, 1975, the 30th anniversary of the German surrender and the end of WWII—which they call the Great Patriotic War—which even today they claim was started when Hitler invaded the USSR in 1941 rather than in 1939 when Hitler invaded Poland and split it with Stalin. Apparently history still is overlooked in Putin's Russia.

The big anniversary was celebrated on *Kurchatov*, and we three anglophone guests were treated to an English language war documentary. The movie made no mention of the Lend-Lease Program and American aid to Britain and the USSR, and scarcely, if at all, mentioned D-Day. Although I was trying to appreciate their huge losses, I was nonetheless pissed. However, when we in the US tout the Normandy invasion, we tend to ignore the fact that most of the German army was on the Eastern Front, not near Normandy.

As *Kurchatov* approached Dakar, we had to turn in passports, as usual on all ships discharging folks in foreign ports. But hours after turning in my passport, I sensed that Soviet peers, including some I had befriended, were avoiding, even shunning me. Could it be that word had circulated about my red, not green, passport and that I worked for the US Navy? Or that I had a German surname and thus German ancestors? Or did I imagine this?

I learned that in Russian, *natsionalnost*, ethnicity, is clearly distinguished from *grazhdanstvo* which means citizenship.

However the next day, docked in Dakar, I sensed no problem. I joined a few on shore leave, mindful that none of them, unlike me, were allowed to be alone.

We stopped to buy some African wooden masks to take home as souvenirs. One of the Russians suggested we only speak English because they try to charge Russians more. I recalled that the Soviet fishing fleet often docks in Dakar.

I said okay, so we spoke in English, not sure if that helped. Strange: I think everywhere in the world 'rich' Americans are likely to be ripped off the most in foreign ports.

I wished these Russians *dosvidaniya*, which means 'til we meet again.' (Indeed I did meet a couple *Kurchatov* scientists again five years after the USSR was no more, on the 1996 *Logachev* expedition to the Haakon Mosby mud volcano.)

We had a small passel of Soviet volcanologists onboard *Kurchatov* who hoped to land on three South Atlantic islands to inspect the geology and collect rock samples. Whether any of them were also collecting intelligence I had no way of knowing.

From north to south were Ascension (8 degrees south), St. Helena (16 degrees south) and Tristan da Cunha (37 degrees south). Together these islands comprise a British Overseas Territory, crumbs still left of the vast British Empire of yore, on which the sun allegedly never set.

There would be no landing on Ascension. The UK, which had a Royal Air Force base on the island, and/or the US, which operated a tracking station for rocket launches from Cape Canaveral, said no. We passed west of the island in the early evening. I remember standing along a port railing with Russian geologists, looking at the bright lights of the RAF base only a few miles away. They looked at me and one asked if I had any influence to get them permission. I thought about saying, "Are you kidding?" But I likely said, "Sorry."

I would visit Ascension nine years later when our Naval Research Lab P-3 measured gravity anomalies over the surrounding ocean, as described in a separate chapter. In 1982, when Argentina invaded the Falkland Islands and was expelled, the British used their RAF base on Ascension. (I published a letter in the *Washington Post* then praising the UK, but wondered what happened to our Monroe Doctrine.)

Kurchatov geologists did land on St Helena, taking the vessel's launch ashore to the small capital, Jamestown, a long narrow town along a narrow valley ending at the coast.

I accompanied several Soviet geologists and shared the picnic they had brought from *Kurchatov*. St. Helena Island is volcanic in origin but the volcanoes are considered extinct. Neither Ascension nor St. Helena are on the Mid-Atlantic Ridge plate boundary between the African and South American plates, but they are close enough to suggest some geological relation.

With several others from the ship, I also indulged in a few hours of historical tourism. Napoleon was exiled on St Helena after his second defeat. The estate still flies the French tricolor, although his bones were returned to France long ago. We stood in the bedroom where he slept and years later died. In this tropical marine climate, the window was open and the shades flapped in the Southern trade winds.

Our southernmost landing was Tristan, the most remote

inhabited island on the Earth. The nearest inhabited island is the one we had just visited, 1,500 miles to the north. Well east of the Mid-Atlantic Ridge plate boundary, Tristan rides on the African plate. The nearest port, from which its 250 or so inhabitants are occasionally supplied by ship, is Capetown, 1,730 miles (six days) away.

Tristan da Cunha is basically one 6,700-foot-high, frequently snow-capped, active volcano, its steep grassy slopes grazed by sheep. The human population lives in the capital, a village called Edinburgh, located on the only small level area. An eruption of lava in 1961 led to mandatory evacuation of Tristan's population, to Capetown. Most islanders chose to be returned later.

The island economy is based on agriculture, export of lobsters and, of course, sale to collectors of Tristan coins and stamps. All land is held in common, so don't think you can build your escape house there. Settlers and land speculators are unwelcome.

Once *Kurchatov* was anchored off Edinburgh, our launch set out toward their small artificial harbor, protected from heavy Roaring Forties swells by a volcanic boulder breakwater with a rather narrow entrance. The tall, macho first mate tried in vain on his first attempt to navigate the heavy swells into their harbor. He gave up and returned to *Kurchatov*, deservedly embarrassed.

Islanders then motored out to help carry ashore the Soviet geologists (and us Western geotourists). Their launch looked more like a giant heavy-duty sardine can, but Tristanites depend on supplies thus carried ashore through the commonly heavy swells.

I wandered around Edinburgh and chatted up locals, who welcomed the uncommon sight of visitors. One family had an eight-year-old daughter. What a great exotic penpal for our similarly aged son, I thought, and copied down the address. There is no airfield, so mail exchanges would be months apart. But Erik showed no interest. Who knows what might have transpired? If the stamped envelopes were saved, it might at least have resulted in a small philatelic fortune stateside.

On our way out of the harbor back to the *Kurchatov*, most of the island population sat on their breakwater and waved goodbye. Of course we waved back.

Not long after I returned to my office from Dakar, I learned that my jaunt with our Cold War adversary had not been reported very far up the chain of command. Maybe some Navy Lieutenant Junior Grade thought it was someone's idea of a joke that a US Navy civil servant was on a Soviet ship.

Of course the spooks (what we called intel folks) soon found out and wanted to interview me. They were likely CIA. I showed them a copy of the *Kurchatov* seismic reflection profiles, but they were more interested in the paper. They snipped off a corner for later chemical analysis.

"We put all such bits of information together," I was told.

Of course they were interested in my 8mm movies, especially that mass of aluminum tubes strapped vertically together on one deck. What did I know of those tubes? I said they were US-made tubes donated to the ship for sediment coring while *Kurchatov* briefly stopped in New York to visit the Lamont-Doherty Geological Observatory before crossing the Atlantic to the Canaries. I'm sure the spooks were a bit chagrined and disappointed.

I suggested that it would have been better if they had known about this cruise beforehand and briefed me on what to look for and film or photograph. The gent from the CIA smiled knowingly and disagreed. I understood that he probably was right. In any case, I had free run of the ship and never was I told not to shoot this or that.

Letters Home

The Kurchatov sailed with equatorial and trade winds of the South Atlantic. Boarded Las Palmas and got off at Dakar, Senegal. Stops in St Helena and Tristan da Cunha islands.

April 9, 1975
Postcard from Las Palmas, Gran Canaria
(Canary Islands) where arrived to board ship

Had a hair-raising car trip around Gran Canaria, lots of bad roads, wild semi-desert landscapes; and two Russians & two English along. Ship conditions not too different. Rather friendly & casual atmosphere. May not land S. Atl. Islands because no permission. Big Hi to Erik & Jason.

April 9, 1975
One day from landing on St Helena

Dear Randi,

Since only six ships call at St Helena each year, you may get this while I'm off on the *Challenger*. Or, we can read it together.

First, our visit to St. Helena. Soon after departing Las Palmas it became evident that we had no clearance to visit any of the islands (Tristan, Ascension, and St Helena, all British but with a US base on Ascension). Radio messages to Moscow and London indicated that the Russians had not requested permission early enough, and we heard a 'no' for two of the islands and a 'standby' for Ascension. John Malpas, a lanky English petrologist from St Johns Univ, Newfoundland, agreed to send messages to London asking for reconsideration, which were eventually answered in the negative. I also asked Alex Malahoff (Office of Naval Research) to do something to help out. Malpas also drafted a radio message to the governor of these islands and we got back a message-let us know specifically what you want to do (an eager party of geologists is on board, anxious for samples). Meanwhile we got a final 'no' from London (Foreign Office) and Udintsev changed ship plans to go to Rio de Janeiro, pick up a Brazilian, and visit Trinidad and Martin Vaz in the eastern South Atlantic. But then yesterday the governor said ok for St Helena, and we'll spend all day ashore tomorrow collecting rocks, and perhaps others from the ship will get to do a little sightseeing. 15 local officials have been invited to a party on the *Kurchatov* tomorrow night, after our day ashore. Now, I don't know whether we will also be able to visit Tristan and Ascension, nor whether we'll put into Rio.

If the latter, I will be home a little earlier, provided my ticket from Dakar is exchangeable. I'd be happy to come home earlier, even now! I guess absences from the family fold are getting harder for me, perhaps they will end entirely someday. Shipboard life is shipboard life, and even at 6000 tons and a Russian ship, perhaps more civilized than ours, the enjoyment is limited. Also I feel I should be in the provinces with the legions—in a twist of that!—fighting for our survival in Maryland. [There were serious plans to move NAVOCEANO—US Naval Oceanographic Office—including our GOFAR Global Ocean Floor Analysis and Research group to Mississippi.]

I had one nightmare of walking into Herb's office (Dr. Herbert Eppert, director of the NAVOCEANO division already for some years down there) after the cruise to hear it was all over: the move was on & immediately. Being away makes our GOFAR lab more attractive (at the Randle Cliffs NRL Annex just south of Chesapeake Beach, Calvert County, Maryland) even than actually being there, and I just know that it will be doomsday when I get home. And last spring, the last azaleas, the last dogwoods, and I'm missing them...

I hope you're surviving—more than just surviving!—my absence, but/and I hope I don't have to worry about you while I'm gone—Any kind of worry! Me? You don't have to worry about me. There are women on this ship, only a few of them, our stereotypical Russian mealsack & some rather pretty. But it's all on the up and up, people are casual and friendly, or at least civil, and there is no obvious hanky-panky. It's all more innocent than in the West. The evening movies are often naive, corny, slapstick, light-hearted, what we would show Saturday morning to the kids on TV. The relationship between people of the same or opposite sex is also more innocent; aggressiveness, sexuality, competition, etc. don't seem to be so developed (yet—maybe another 10 years will change things).

People go out for 3 or 4 month cruises with only $120 in foreign currency to spend, but, on the other hand, they don't work as hard (more people shorter watches, for one), and there is more possibility of recreation, partly because of the larger size of the ship: volleyball, chess, ping pong, shuffleboard, and '*Kurchatov* billiards' may be seen on the fantail at any time of day. Since my film won't work in the darker labs, you will see a lot of that including, a Russian Neptune's Day—yes I had to do it again!—when I get home.

The food takes getting used to, no milk, ice cream, almost no cheese, no coffee (only tea, including afternoon tea served with vegetables or salad) and absolutely no sweets. More on all that later.

On the scientific side I'm a little disappointed, not because of their technology isn't up to snuff—it is in most cases, and in any case more diversified—but because my suggestions about what we ought to be doing are not taken very seriously. I'm not having much input to the cruise objectives, and that wipes out a good bit of the scientific viability. Not that other scientific objectives outweigh—they're very casual about that and there is very little discussion by the other Russians, either with the chief scientist Gleb Udintsev about what we should or should not be doing. On the positive side, my Russian is getting much better, except for declensions and conjugations which I once knew and since forgot.

Unfortunately, although I'm the only American, there are two English speakers, John Malpas (my roommate) and Susan Humphreys, a pleasant, intelligent graduate student from Woods Hole. Since neither speaks Russian, it doesn't help mine, and some of the Russian scientists are quite adamant about speaking English to me. Others are amazed that an American tries to speak their language and right away overloads my linguistic circuits. They have had little chance to learn, like we English-speaking people, how to speak a slower, simpler form to be understood by foreigners. Anyway I can talk about faults and lost dredges and ship's speed and sensors and magnetic susceptibility, a rather narrow and elsewhere useless vocabulary, and be understood. But I don't get my grammar corrected, nor do I correct theirs [English], because all one tries for is to get the meaning across. The rest is Polish anyway.

So much for a little bit of what it's like. Erik must be walking on crutches by now, and back in school (he broke his leg on the way up a T-bar while we visited my brother in Switzerland). Also, I hope my planet pageant is progressing.

[I had written a play illustrating the solar system for the Lower School of Calverton, a private school in Calvert County. The science teacher continued working on Pageant of the Planets after I shipped out. The solar system—planets, sun, moons and comets—was represented by ages 1-5 and the 6th graders narrated. Each kid carried a colored balloon with a small light inside, connected by wire to a battery in his or her pocket. The gymnasium floor was turned into the solar system and another parent and I climbed on the roof to cover the skylight with a tarp. I came back in time to see it performed—no comet kids ran into planets or moons.]

Maybe a university faculty position would be better. I don't want to go out on many more cruises. If I leave NAVOCEANO, it will probably be the end of cruises anyway. This one is 'once in a lifetime,' and I think it will have been worthwhile. I hope you get this sometime soon.

April 15, 1975
Akademik Kurchatov. About 8° W, 26° S

Dear Randi,

When will you get this letter, if ever? We shall see. Our plans seem to be ever fluxing, a situation that helps alleviate the boredom of being merely a visiting scientist. As of last night, we are on our way to Tristan first, and then the main thrust of the expedition, an investigation of the Walvis Ridge (a large sub-marine ridge between southern Africa and the Mid-Atlantic Ridge). Gleb Udintsev would rather not visit Tristan, but he has a small horde of eager petrologists, including one influential in the (Soviet) Academy of Science, to contend with. They joined the ship in Dakar (Senegal) and will disembark when I do. The petrologists have not been supplicated with dredge results. On the two seamounts and a knoll west of St Helena, one was lost, two came up empty, one gave only extremely altered basalt, several had manganese crust, limestone, corals & shells, but one however yielded an unusual assortment of large, detached plagioclase crystals the size of your thumb. The Russians are willing to share their samples with us foreigners, by the way more so than a lot of US geologists would.

Presently heading for Tristan under cloudy skies and fairly smooth seas. Yesterday we were pounding into a moderate SE trade wind sea and I felt somewhat rotten—not at all helped on hearing *Kurchatov* will not arrive in Dakar until 6 April [must have meant 6 June]. When the date is secure I will send you a loving radiogram.

[Personal mush omitted]

I had hoped to write papers etc but so far all I've done is write out a seminar in Russian, which Misha (Dr. Michael) Kogan the gravity man, and a broad intellectual, helped correct. Although I (separately or with John Malpas or Susan Humphreys) have spent a number of afternoons or evenings talking, drinking, and snacking on sometimes weird combinations of things, Misha Kogan's gravity lab has been especially hospitable. I will probably give my Russian hotspot seminar tomorrow—another 'first' for me—and maybe think about making it into a short review article, as Misha suggested.

By the way, some people do get drunk on this ship—I have seen both my roommates, and *Kurchatov's* middle-aged, goateed, sea-floor photographer in this condition.

[Omitted mush followed by career counseling]

I have complete confidence that you will find pride and meaning in the post-baby years ahead! You're young and intelligent and imaginative, and you have choices, maybe you have a few—nevertheless valuable—false starts before you find something enjoyable and challenging. [In later years Randi got an MA in Urban Planning, becoming a professional planner for Calvert County Dept. of Planning and Zoning; she introduced the concept of Transferable Development Rights to the county, which became the first of many US counties to adopt this technique to preserve land without use of public funds.]

I don't want to write too much because there is no assurance this letter will ever be mailed. In the first place, Tristan is exposed to the late-fall sub-Antarctic seas on all sides, and if the seas are too high no landing will be attempted. If it is, I may not be invited as I am no petrologist. (But I'm going to try like hell.)

We will be there the day after tomorrow, or a day later. The total population is about 200, who live in the only 'town'—called The Settlement or Edinburgh. Tristan is a single towering volcano surrounded almost on all sides by formidable sea cliffs. It might be the most isolated permanent settlement on our planet. Well, you'll hear all about it later, if I'm lucky enough to get to go.

A Soviet (Estonian) Scientist in our Kitchen

After a brief stopover in Calvert County, I left home again to join the drill ship Glomar Challenger *for July and August, 1975. Now that two-month voyage was coming to an end, and I was about to rejoin my family after another long absence.*

The *Glomar Challenger* was approaching the US East Coast and its next port stop, Norfolk, Virginia. Only two days more and I would be disembarking for home. I sent Randi a telegram from the vessel, and she sent one back from her and the two boys, Erik and Jason. My family would drive down from Calvert County and meet me at the dock. As we approached the dock I climbed to the top of the drilling derrick. There was my family! I waved down and they waved up!

As we waited for clearance to disembark, I said goodbye to our team of scientists, most of whom I would never see again. When I shook the hands of our shipboard sedimentologist, a Soviet Estonian named Dr. Ivar Murdmaa, I learned he had just (by radio message) been ordered by Moscow to remain on the vessel for another two-month leg. He could not even get off the vessel for this port stop because Norfolk, a major US military facility, was off limits for Soviets. The port step ending Leg 43 had been changed to Norfolk while we were at sea. This was, of course, Cold War tit for tat; certain Russian cities were closed to Americans.

I felt sorry for his predicament and made a decision that could have at least cost me my job. I invited him to spend his port stop time with my family and me in Calvert County, which was not off limits—I was sure. He would remain in my car as we drove through Norfolk and we would make sure he got back to the ship in time. The cognizant official on *Glomar Challenger* did not object, but I did not check with my Naval Research Laboratory office. Why get bureaucrats involved?

Ivar Murdmaa was born in August 1931, in Estonia, and died in December 2023. He began his career as a sedimentary geologist and spent most of it in Moscow as a researcher at the Institute of Oceanology at the Russian Academy of Sciences, where he became head of the Laboratory of Mineral Resources. Murdmaa worked on bottom sediments in the Arctic Barents Sea and became the first Estonian to visit Antarctica while working for the Soviet Antarctic Expedition. He founded the Estonian Geological Society.

On the way home, I passed the usual blackboard sign on the deck showing departure date and time. I would not be on that next leg. Once down on the dock, I hugged my family and explained the situation with Murdmaa to my wife Randi. She was totally supportive of the invitation. Estonia was just across the Baltic from Scandinavia.

Our car trip to Southern Maryland from Norfolk is mostly through scenic, historic, and rural tidewater Virginia. It must have been a bit cramped for three adults and two small kids in a Volkswagen Rabbit, but I imagine he was used to crowded small cars in the USSR. This might have been Ivar's first visit to the US, and he was a great houseguest, interested in many things. He was fascinated by TV commercials and taught our 7-plus year-old, Erik, how to make paper airplanes. Randi took him to a major department store and to an art gallery. From there he phoned his Soviet embassy in DC. She watched our Estonian carefully and recalls how nervous he looked, even shaking. Too bad we couldn't hear the other end of that conversation.

Was it *Defect and we go after your family*? Of course he never expressed any hint of wanting that.

On his last day, we dropped him off at the DC Trailways bus station for his trip back to Norfolk and the *Glomar Challenger*. We considered first stopping downtown for a restaurant meal and possibly visiting a museum. But we were tired and decided to head directly home. Sometimes a minor decision can have really major consequences.

We had scarcely opened the door when the phone rang. The voice on the line was that of a GOFAR colleague, Troy Holcombe. Troy was the most imperturbable, unshakably calm person I ever met. But on the phone he was speaking in an excited, raised voice. *Where is Murdmaa—the* Challenger *is ready to get underway.* Both Murdmaa and I had misremembered the departure time. It was not tomorrow but today. Now.

Holcombe told me to hang up and call Melvin Peterson, the Scripps head of the Deep Sea Drilling Program. I phoned Peterson, whose voice was also excited. I started to apologize but Peterson interrupted. Explain this all later but just tell me where he is right now. I told him the Trailways bus departure time and hung up.

What happened next was this: The ship sent a car up the Trailways bus route to intercept that bus, get Ivar into the car and back to the ship. Amazingly, the *Challenger* waited for him to embark. I can just imagine the fright of a Soviet when, at the bus stop, someone probably unknown to him stepped up next to the driver and asked: "Is there an Ivar Murdmaa on this bus?"

Had the vessel sailed without him, it would have been an international incident and almost certainly cost me my job. I heard nothing about, or from, Ivar. A few years later, he sent a Soviet-type Christmas card—with Father Frost instead of Santa Claus. I don't remember if he wrote anything but I trust he had fond memories of his visit, and no anger about me forgetting the departure time—which of course he had also forgotten.

Murdmaa's dreams about the end of Soviet communism, the breakup of the USSR and the restoration of Estonian independence (lost in a violent 1940 coup) came true. His biographer recalled standing next to him, protesting Soviet tanks in Tallinn as Gorbachev tried to quell the revolts. However, we deliberately never brought up politics on his visit.

During the course of his long life, Murdmaa developed diverse interests and broad horizons. He wrote travel books and non-fiction articles and became a leading environmentalist. His interest in art—shown by his wish to view Salvador Dali paintings while visiting us—had likely been inspired by his artist father, whose other three children became successful as an architect, ballet master, and physical chemist. He was known as a great conversationalist, a trait that shone through language issues on his visit.

Reading his obituary made me regret not having reconnected with him—especially in the 1990s, when with my US colleagues I collaborated with Russian geoscientists and visited Moscow, St Petersburg, and Arkhangelsk. We could have compared our common interests in the environment, writing, travel, and even Antarctica.

And we would have recalled his 1975 visit, when both of us nearly landed in trouble.

CHAPTER 3

Voyage to Nowhere On the Russia's Mendeleev

The Cold War end and collapse of the USSR opened new opportunities for applied marine science. In this case the world needed to know more about the nuclear waste dumped into the ocean by the Soviets. It was about assessing hazards and the environment.

Not sure what was known then; today, thanks to the International Atomic Energy Agency (IAEA) we know that the USSR dumped more than twice the rad-waste—hazardous waste that contains radioactive material—of all other nations from 1946 to 1982.

High-level rad-waste dumping was banned by the IAEA already in the 1950s, and low-level waste by the 1983 London Dumping Convention. Waste radioactivity had declined by 1993, but much remained, owing to long half- lives.

US funding sources gave us at the Naval Research Laboratory an opportunity. The Lab is not funded by appropriation, as are most other US government agencies. We had to, as they say, sing for our suppers. An early step was attending a nuclear waste conference in Arctic Russia: Arkhangelsk.

Environmentalist organizations like Greenpeace were also there and, in my opinion, well-intentioned, but not well educated in radionuclides or ocean geology.

Human society is full of competition and/or collaboration—between individuals or between nations—or even between government agencies. Administratively, our Lab is under the Office of Naval Research. In this case, we at the Naval Research Lab initially were competing with our parent agency, the Office of Naval Research. Eventually, we began cooperating.

Our plan was to charter a Russian research ship and help fund their scientists who, with our US team, would investigate the Kara Sea. It was there, in a seafloor trench just east of the Novaya Zemlya archipelago, that the USSR had dumped lots of nuclear waste. We know today from Radio Free Liberty that the Soviets dumped about 17,000 radioactive objects into eight areas—totaling some thousands of square miles—of the East Novaya Zemlya Trench. This includes some 18 nuclear reactors and one entire, severely contaminated submarine, the K-27, scuttled in 1982.

In 1993, with others, I assembled a team from the Naval Research Laboratory, the US Department of Energy laboratories, the US National Lab, the Naval Academy, and elsewhere. A small group of us flew first to Moscow to arrange the deal. I got the impression that folks from the Navy's Naval Reactors office were not thrilled about this project. They were, I think, worried that our snooping into Soviet waste might lead to reciprocal snooping into US rad-waste and lost nuclear submarines.

Our interagency team converged in Kiel, Germany, to await the scheduled arrival of *Dmitry Mendeleev*, a *Kurchatov*-class research vessel. I was probably the only team member who had been on such a ship—on the 1975 South Atlantic *Kurchatov* expedition 18 years earlier. As well, I was the only team member somewhat fluent in Russian. Our vessel would be coming from Kaliningrad—the pre-WWII East Prussian capital of Koenigsberg—from which Germans were expelled and replaced with Russians as part of the largest ethnic cleansing operation in modern history.

The *Mendeleev* was operated by the Shirshov Institute of Oceanology in Moscow. About 400 feet long and 5,500 gross tons, the Soviet era vessel was designed for roughly a 47-member crew, along with 17 officers, and 78 scientists. It had a large lab and cargo space, plenty big for our mission to the Kara Sea.

Our team members, scattered among several Kiel hotels, got acquainted with one another over good German beer. The *Mendeleev* did not arrive on time. There was one delay after another. Our team was running up hotel bills, ultimately paid by the US taxpayer. Of course the beers were paid for by our own Deutsche Marks.

Finally the Russian vessel arrived and tied up along Kiel dock. When port formalities were done and the gang plank lowered, I, as expedition leader, was welcomed aboard. I greeted a few whom I had met on the *Kurchatov*. As typical in Russian culture, I was immediately ushered into a cabin for some shots of vodka, with toasts to a successful cruise. After these toasts, the Russians' chief scientists mentioned there was just a small problem but not one that should not handicap the mission. The Russians took me into their chart room, where a map of the Kara Sea lay spread out. The map, showing geography and water depths, had a single straight east-west line penciled on it. The line ended at one point in the East Novaya Trench.

We were permitted to follow that track and investigate only that one point.

When I was assured this was not a joke, I said it was a non-starter but that I would go back to my hotel and phone my superiors in DC. I was assured that my concerns would be relayed to Russian authorities. In the meantime, my various teams would start to load their gear onboard with the ship's crane. The Russians—marine geologists—implied that much new geology would be discovered and that this would justify my approving this very restricted access. I assured them that as geologists we would welcome all that—but that this cruise was funded to map and test nuclear waste.

I next spent days and days in my Kiel hotel room on the phone with various US government authorities, both from the Navy and the State Department. To a large extent I, as expedition leader, was trusted to make my own judgment. Members of my team were getting tired of wasting time waiting. So, given that the Russian authorities—not the scientists and officers on *Mendeleev*—offered no compromise, I opted to abort the mission. The US gear already onboard and partly installed was now off-loaded again.

The German dock workers were puzzled. Did the Americans and Russians have a falling out? I tried to explain.

After some months, the funds sent to hire the *Mendeleev* were returned to the US. Not, however, the many days of room and board in Kiel, and the travel costs for our team.

This debacle seems to have been created by some ex-Soviet admiral who was pissed that the US was taking advantage of the USSR collapse to snoop into matters related to the Russian nuclear know-how. Would the US DOD object to Russia investigating the *Briggs* with its nerve gas containers? Or the two lost US nuclear submarines?

In the meantime, US nuclear waste disposal remains contentious. Nevada does not want nuclear power plant waste stored at Yucca Flats. In Southern Maryland, low-level nuclear waste stored in dry casks continues to accumulate at the Calvert Cliffs Nuclear Power Plant, where one unit is licensed to operate only until 2034 and a second until 2036.

The Kara Sea rad-waste we had planned to investigate is still there today, three decades later. Although some of the radio-activity has naturally decreased, the mess remains hazardous. Prodded by Norway, Sweden, and others, the Russian nuclear agency, Rosatom, in 2020 issued an ukaz (decree) to oversee the cleanup of the most dangerous materials during the period 2020–2029.

In all probability, the war Russia launched in Ukraine delayed this plan.

CHAPTER 4

Professor Hauls Boatload of Russians to Mud Volcano and Rift Valley

Our 1996 expedition on board the Russian Federation's *Professor Logachev* was science-wise the top of my Russian collaborations. As also for the *Mendeleev* and *Keldysh*-Mir ventures it was mostly arranged by WHOI and my colleague Kathy Crane, who had previously collaborated with Russian ocean geologists. Her international connections and expertise were vital. At the time, I had arranged for her to work at NRL, sort of a loan from academia.

The expedition included a small Norwegian team led by Eirik Sundvor of the University of Bergen. My NRL team and I flew to Bergen to await the *Logachev's* arrival from home base Shirshov Institute of Oceanology (St. Petersburg). The Russian co-chief scientist was Georgi (Gosha) Cherkachev.

The vessel was late and its skipper skipped the Norwegian pilot—contrary to rules, perhaps to save $$—but managed to navigate the fjords safely to Bergen.

We brought a number of Russian scientists to the Haakon Mosby mud volcano where the teams deployed deep-tow side-scan sonar, bottom photography, heat flow probes, and sediment sampling to map the mud volcano in detail. The sampling and imaging proved this was an active mud volcano, supporting a vibrant "cold seep" chemosynthetic ecosystem.

The conspicuous white areas we thought might be gas hydrate were eventually recognized as bacteria mats.

From the mud volcano we headed north to the Knipovich Ridge rift valley which forms the boundary between the slowly separating North American and Eurasian tectonic plates.

Various types of lava flows were photographed, some only slightly dusted with sediment so evidently erupted rather recently. Bottom water temperatures and mineral coatings hinted at nearby hydrothermal venting, but no actual vents were discovered.

We had several parties attended by those not on watch, and some scientists gave seminar-type lectures. I had invited Jeff Halka from the Maryland Geological Survey to join the cruise as guest geologist and he pitched right in. Jeff's talk was about the Chesapeake Bay and like other English-language lectures was translated into Russian in real time for attendees knowing little English. The interpreter was the daughter of a noted Russian Arctic Ocean geologist whom I had cited in my PhD thesis but never met.

When it became my turn I lectured about hotspots but in Russian. Our interpreter duly translated my talk into English for the Americans and Norwegians in the audience.

One of my team was the NRL historian and archivist David van Keuren. Just before some of the scientists disembarked in Longyearbyen (Spitsbergen) to fly home, van Keuren interviewed Cherkachev, Crane, and me. The transcript of that interview is an Appendix.

After the cruise we hosted a small international party in our cottage in Scientists Cliffs. Sherrod Sturrock attended as non-scientist but was inspired to write a poem about mud volcano with tubeworms.

CHAPTER 5

In a Russian Submersible, Below the Nordic Seas

I didn't eat or drink anything the day before, then scarcely slept a wink, tossing and turning with apprehension.

It was early July 1998, and I was out on the Norwegian Sea on the Russian ship *Mstyslav Keldysh*. Up where we were, it never really gets dark on a July night. It was already light when a small elderly woman in a blue jumpsuit entered my room upon first knocking. The time had come.

With Mir, splashing meant a crane dropping us in the drink, at which point we would begin bobbing and rolling before sinking into the deep.

She handed me what I was supposed to wear. Also blue. I had heard she was Russian so I said, *Dobroye utro* and *spasibo do skorovo* as the woman left. She smiled slightly and nodded. It was clear she had done this countless times but probably was surprised that I knew any Russian. She had met many Americans before, but very few who knew her language.

The sea state was light, so I didn't have to hold onto my bunk's roll bar as I struggled into my blue jumpsuit, emblazoned with colorful official insignia.

This was not about any operation, at least the hospital kind. I was about to board one of the two Russian Mir submersibles and spend 10 hours under the surface of the Norwegian Sea. It would be my first submersible dive. Manned submersibles have no bathroom facilities. A bottle in case of No. 1, while seated next to the pilot and engineer. In the event of No. 2? God forbid.

It had only been only a few years since the *Titanic* filming ended, explaining why that blue lady and most of the *Keldysh* crew had met many Americans. I might have had the cabin of James Cameron, the Canadian deep-sea explorer and director of the *Titanic* film.

Indeed, as one of four chief scientists now researching geology, not shipwrecks, I had an uppermost cabin. The others were German and Norwegian, and we'd been strangers until our introduction at Tivoli Gardens, the Danish ancestor of Disneyland.

On our transit from Copenhagen to the dive areas, Anatoli Sagalevitch—the skilled submersible pilot who was also the colorful director of the Mir program—sometimes complained about his and Mir's part in the *Titanic* movie. His two gripes: Cameron should have paid the Russians a lot more, particularly after the box office success of the film. Secondly, Anatoli was disappointed in having takes of his ukulele Russian folk songs wind up on the cutting-room floor. He appears only briefly, looking sporting in his blue dive suit.

Dressed for dropping into the deep.

Outfitted proudly in my own blue diver jumpsuit, I headed for the starboard boat deck near Mir 1. Mir 2 was sitting in the same position on the port side. It would be launched soon after ours. Just before climbing aboard the three of us, pilot, engineer and me, were photographed.

Besides Russia, the US, France, and Japan all have or had manned, deep-sea manned submersibles with pressure hulls strong enough to withstand the enormous pressures at depths of many thousands of feet without imploding. Deployed throughout the world's ocean for decades, mostly for research, none of them ever imploded—as the commercial *Titan* did in 2023.

The Russian system is unique because the Mir twins can operate at the same time and, should the need arise, one Mir could rescue the other. Cameron contracted with the Russians for the *Titanic* filming not only because of lower cost but also because photography of the shipwreck is so much richer if there is side-lighting. One Mir illuminated the wreck from one side, and the other captured the video.

I climbed up a ladder to the submersible top. A crew member standing up there took my shoes and handed me a towel and my own digital camera, an audio recorder and a mic to record what I saw and conclusions I might reach. Of course, the Mirs had plenty of cameras operated by the pilot or engineer. Diver scientist/observers were not allowed to operate those or touch the various joysticks.

I had assignments other than just observing seafloor geology through my starboard porthole. I was to measure the temperature gradient in the mud with a rod-shaped device designed by my colleague, Kathy Crane, who also was a diver. She had instructed me on the experiment I was supposed to perform. In addition, I got a cram session on fish to look out for, report on and, ideally, photograph. The ultimate prize would have been glimpsing a rare Greenland shark.

Our expedition fish expert was an American based in South Africa. The Russians also had marine biologists on board.

There were far more scientists on board who said they wanted to dive than there were slots.

With the pilot, engineer and me in our seats—mine was a padded bench—the hatch was closed and sealed. Our 10-plus-hour trip had started. Our pilot talked to the shipboard control. Then all could feel the movements as Mir 1—no bigger than a short bus or long van—was lifted up and over by a crane, then down into the ocean.

I felt the instant our submersible touched down into the ocean. We began to move up and down and roll. The next instant all we could see from our viewing ports was churning bubbly water. Then the support boat arrived, with its crew of three in wetsuits. We could hear some commotion on top of our Mir as one of what the dive team called 'cowboys' stood above us and attached us to the towing cable while detaching us from the cable connecting us to our mothership, the *Keldysh*.

Then we were on our way, safely distant from *Keldysh*. The support boat approached again, unseen from Mir, and the cowboy jumped on top, unhooked us, and jumped back into the boat. We were on our own. I watched our submersible pilot, a short, tough former MiG pilot, as he contemplated the large panel of switches and buttons and gauges. As a passenger in small aircraft flown by friends, I know they first take out a checklist before touching any controls. I expected our submersible pilot to do the same.

I was aghast when he did no such thing but started flipping switches and pushing buttons. Reason told me he was experienced and trusted his memory. Russians are not suicidal, although I knew from my past experience on their research ships (1975 and 1996) they are much more cavalier about safety than we Americans.

We began our one-hour descent to the ocean floor nearly a mile below. Gradually, the rocking and rolling diminished and there were fewer bubbles in the water outside. It got darker outside my porthole.

Then total darkness outside. I hoped the pilot would turn on the floodlights so I could look for strange fish and other creatures. However, he did not. Much to my disappointment he reminded me in broken English, or perhaps in Russian—which I somewhat understood—that this dive was for geology and life on the ocean floor. The Mirs have to conserve battery power.

As we finally approached the sea floor, according to their echo sounder, I began to stare into my port expectantly. My face close to the thick glass, my view became obscured by condensation. The water outside was not much above freezing. I now realized why the lady who took my shoes before I climbed down into the submersible had handed me a towel. It was not about leaks; any leak would have spelled imminent implosion, as happened on *Titan* on June 8, 2023. The towel was to wipe condensation from my side of the port.

The pilot switched on the powerful floodlight. Outside my port was swirling marine snow—suspended sediment and microorganisms, dead and alive. Predators and prey. Finally there was a faint brownish tint below us. The ocean bottom itself!

I had been researching the seafloor and below from the surface off and on for the last 34 years. Some of my colleagues and peers above us aboard *Keldysh*, among them Kathy Crane, were experienced submersible divers. Now I was seeing the bottom for the first time. I had my 1998 digital camera ready for the ocean floor, but it was just dull brownish mud.

The view to the port was limited by equipment outside to be manipulated by the pilot. But wait; what is THAT off some distance to the starboard? White and ghostly, rising above the seafloor like some flower ghost. Although not a biologist, I realized it was an animal called crinoid, an invertebrate sporting what looked to be a white feather duster for capturing prey, rising in the currents.

This strange creature also resembled fireworks exploding in the sky. How appropriate; today, my first dive in a submersible was, in fact, the Fourth of July. I captured that crinoid with my digital camera and recorded observations in the mic around my neck.

The mission of the dive was to examine the sidewall of a humongous sub-marine landslide that caused tsunamis along Norwegian and Scottish coasts during the late Stone Age, some 8,000 years ago. We looked for vents where methane might be escaping or other signs of recent seafloor action. However, all looked peaceful. We moved slowly along the seafloor at a knot or so, careful not to churn up clouds of view-obscuring mud with our propellers.

At one point I noticed a small outcrop of hard sediment layers, old sediments exposed in the sidewall by that giant slide. I have no Mir photograph of that or anything from that dive because I had misunderstood. No photographs were permitted unless the scientist observer requests one.

I had assumed there was continuous video and pilots decided on stills. My bad. Fortunately we didn't happen on to anything important for our mission. Well, I did immortalize one crinoid and some more mundane scenes with my *Keldysh*-issue camera.

Views were always limited by ubiquitous marine snow. Maybe not as bad as the Chesapeake in summer or after heavy rains. But even in the clearest water, one can't see very far. Water scatters light in all directions depending on wavelength. Blue-green is the best wavelength, but still very limited. This is why sound, not light, lets humans probe the oceans except for really up close and personal.

As we crossed a small, flat, mud-covered plain, I motioned the pilot to stop and set our Mir gently on the seafloor. I chose this spot to measure mud temperature from the sea floor down 50 cm. This was the length of colleague Kathy Crane's probe with thermistors at several places along its length. How fast Earth warms with depth tells scientists something about what lies below. I had a small laptop and connected it to a Mir outlet.

Meanwhile, the pilot, using joysticks, pulled the probe out of its holder outside the Mir and manipulated it away and then into the mud. We waited some minutes; then I downloaded the readings to my laptop. The probe was pulled out and placed back in its holder. Mission accomplished!

We were not alone in these depths. No, I don't know about any silent submarines between us and *Keldysh*, but I do know about the other submersible, Mir-2.

On most missions, including the *Titanic* filming, they are down at the same time. And they talk to each other and to

Keldysh via acoustic phones. Hearing the other Mir was spooky, as if in a giant cave. "*Mir odin miro dwa,*" the hollow voice intoned in Russian. Of course. *Mir 1 (this is) Mir 2.*

We were well into the dive when something I couldn't have imagined happened, down here 4,000 feet and several hundred miles from land. A standard white plastic bag lazily drifted into the manipulator gear. Or we moved into it. The bag got itself stuck and the pilot had no means to get this cursed plastic bag off, so it obscured much of my view for a long time. Welcome to the global age of plastic beneath the sea.

Peter and two US colleagues chat with Anatoli Sagalevitch,
head of Mir program, at Tivoli (Copenhagen).

At the end of our day, we parked on the seafloor and enjoyed a Russian picnic. Some of the fare was near the lever that could release lead weights and allow the Mir to rise to the surface buoyantly in case of total electric power failure. We would be back on *Keldysh* in an hour so it was safe to eat and drink.

My day on the ocean floor was in one way like a typical eight-hour day in my office at the Navy Research Lab, with an hour commute one way.

My second dive more than a week later was very different in several ways. Knowing what the experience was all about, I was no longer apprehensive. On the flip side, I also recall less. My second dive was for that time perhaps the northernmost submersible dive. The location was in the Fram Strait between Spitsbergen and Greenland. Working with the Norwegians on the University of Bergen ship *Haakon Mosby* in 1989–'90, we had discovered evidence for methane venting on a seafloor sediment ridge. Now from *Keldysh* our team would inspect these vents up close.

From my port, I could see white sulfur bacteria mats, starfish and other life in this chemosynthetic ecosystem, so unlike the photosynthetic (plant-based) ones we know well from land. A type of small bottom fish called eelpout is supported by this strange ecosystem (and would no doubt taste terrible). The shipboard ichthyologists wanted samples, so I pointed to a fish and the pilot maneuvered his slurp-gun nozzle near the unsuspecting fish and vacuumed it into a container with one of his joysticks.

Several dives took Mir scientists down to the Haakon Mosby mud volcano, discovered on that ship (and later named after it) in 1990, verified by sediment cores in 1995 and more extensively on the Russian ship *Logachev* in 1996. (Read more in the *Haakon Mosby* and *Logachev* chapter with its appendix.)

Other Mir dives examined the Knipovich Ridge plate boundary between the North American and Eurasian tectonic plates. Different kinds of basalt lavas were observed, some only lightly dusted with sediment and thus erupted not long before. There was the long-shot hope of discovering a hydrothermal vent with hot water rising out of the bottom, and perhaps strand shrimp, clams, or tube worms. As for most long shots, nothing was found. Hydrothermal vents were known to be very local and widely scattered along plate boundaries where plates move apart.

We four chief scientists—leading the US, German, Russian and Norwegian teams—had agreed to let a small team of Russian nuclear experts come along and take one dive down to the wreck of the Soviet nuclear attack submarine *Komsomolets* (K-278), which caught fire and sank April 7, 1989, with the loss of 42 lives. This wreck lay on the lower Barents Sea continental margin close to our diving targets. The Russian nuclear team, which had boarded on our Longyearbyen (Spitsbergen) port stop, wanted to check on possible leaks of radioisotopes, notably plutonium from the torpedoes. Obviously, none of us were invited to participate or to ask for dive results. Once *Keldysh* had stopped 5,600 feet above the wreck, the Russians had a brief ceremony in which they remembered the loss of life; a small wreath was launched overboard in silent remembrance. We scientists observed in respectful silence. We later learned that no new leaks were detected.

Also on board were folks from National Geographic, which planned a TV piece about our expedition. The teenage son of photographer Emory Kristof got to come along. Kristof had photographed the *Titanic*. I can't imagine this would ever be allowed on any US academic or government research vessel. Incredibly, the program they later produced said little about the science of our dives but focussed on this teenager's birthday celebration on board *Keldysh*. The video did include the two misadventures.

Our project's last dive—we were behind schedule—went to my friend Eirik Sundvor, a University of Bergen professor. He would inspect the bottom of the Molloy Deep, at 5,550 meters the deepest spot in the Arctic. Between Spitsbergen and Greenland, the Molloy Deep is also the plate boundary, a place where the two great plates scrape past each other. Sundvors' trip three miles down and the return three miles up used up most of the allowed time. The other Mir was not launched. The dive location was dangerously close to the edge of the Arctic sea ice edge, dramatically shown on the bridge radar. Big trouble if the wind were to change and move the pack over the Mir recovery site. Moreover *Keldysh* lacked an ice-reinforced hull.

Sundvors' dive also found that spot on the floor of the Molloy Deep geologically boring. But this could not have been predicted. Many spots on the Moon and Mars don't shed much light on geological mysteries either.

The Molloy Deep dive was not the riskiest and potentially most calamitous event on our expedition. There were two events that might have had unhappy endings. My colleague Kathy Crane and Russian co-chief scientist Georgi (Gosha) Cherkashov investigated the Haakon Mosby mud volcano on Mir-2. Anatoly Sagalevitch was down there as well on Mir-1 but got the submersible stuck in the mud. Mir-2 came to help, but its manipulator arm was broken.

Fortunately the head of the Mir program and the most experienced diver managed to free the submersible by pumping ballast back and forth and between tanks, rocking Mir-1 loose.

The other near-calamity story began when I happened to visit the *Keldysh* meteorologist, a kindly older lady whose English was worse than my Russian. I do okay in that language if the topic is science. There are many terms similar to English, with the same Greco-Latin roots.

She spread out her large paper map, the kind with isobars, wind vectors, etc. A low pressure system was forming and intensifying. She had recommended against dives the next day but Anatoly, the boss, would hear none of it. I suspect he wanted to give his western paying clients all the dives we had planned. Maybe a bit of bravado, as if to say we Russians are no wimps.

The Mirs had never previously operated this far north of the temperate and tropical venues in which they were used to operating. The Greenland-Norwegian Sea is notorious for relatively small but rapidly intensifying storm systems. Both Mirs were launched in somewhat choppy seas with no problems. As winds picked up during the day, Anatoly ordered both vessels to shorten their time on the bottom. By the time Mir-1 was recovered, the seas were very rough. But the submersible was safely hoisted up.

It had grown even stormier by the time Mir-2 rose to the surface. The vessel was being tossed around, as was the support boat carrying the cowboys, and all this was happening in plain view of *Keldysh* crew and scientists not on watch.

We on board were holding onto railings, but I was able to take photos. We could see the cowboy in wetsuit jumping onto the top of the Mir to attach the cable only to be washed into the ocean by the next sea. It seemed like forever before this exhausted man succeeded.

Once Mir-2 had been towed near *Keldysh*, the next step was even harder. The other cowboy had to disconnect one cable and connect the one dangling and swaying from the crane. Over and over, he was washed back into the sea before he could do the job.

The recovery boat tried hoving to for some time, and the cowboys were given a rest. I think Anatoly considered having Mir dive again and wait for the storm to move on. However Mirs have only so many hours of oxygen beneath the surface. Finally, he decided to have the cowboys try again, and after being swept into the sea again, the Russian cowboy succeeded. The crowd on *Keldysh* cheered.

The unfortunate scientists on Mir-2 were Kathy Crane and the ichthyologist from Capetown. They were repeatedly dumped into each other's laps as the Mir bobbed and rolled for hours. I don't recall any seatbelts. I do recall that Anatoly later regretted not listening to his weather forecaster.

My takeaways from this expedition are complex. For me, seeing the ocean floor in person was a powerful and memorable experience. However the same seafloor could have been viewed more cheaply and safely using a deep-towed video or a UAV (Unmanned Autonomous Vehicle). Seeing a video of a thing or a human is absolutely not like seeing it in person, but it's hard to quantify the distinction.

This issue closely parallels the space program. Sending astronauts to the Moon or even to Mars is far more costly and challenging than sending advanced rovers. Even so, the desire to explore in person and then report the experience is deeply ingrained in human nature.

Letters Home

June 24 1998; 9:30 AM
from Copenhagen

Dear Randi,

Sunny evening after clouds. A bit seedy, more paper in the streets including 7-Eleven & McDonalds wrappers. Woke up briefly at midnight last night & heard St Hans fireworks at nearby Tivoli. (Haven't been there, but took a train to Roskilde & Viking ship museum you & I saw in '93, now expanded, hand-on craftsmen building ships outside & people can take sailing courses on Viking boat replicas, oars & sail & all!

Most of our stuff is already at the agent's. Ran into two of our party on the street. *Keldysh* sailed from Russia a few hours ago, docks tomorrow night. I will visit our ichthyologist at the zoological museum where he is working tomorrow, then on to ship. I bought some books on Vikings for your work on Jay's (Jacob Stampen, my brother-in-law) birthday. Hope I have time to send them.

June 27 1998

Hi! Another big scare! The NRL check had been misplaced for several days in a Danish bank & was only found 2 hrs before bank closing Friday. We are set to sail 2000 tonight. Just about everyone has assembled, labs & staterooms assigned, equipment loaded aboard but not installed, a 'last supper' at Tivoli's courtesy of National Geographic. I will try to get a hot water heater for my instant coffee after mailing this, gotta check out of hotel [obliterated by postal tape] luck! Peter

In the 1970s, I collaborated with paleontologist-turned-cartoonist
Jack Holden, who produced complex art such as this depiction
of a scientific argument in which I landed. The scientists standing on Earth's
core from left to right: W. Jason Morgan (Princeton), Don Anderson
(Caltech) and Gillian Foulger (Durham University).

PART V

Afterword: of Poems, Parody, Big Plans, and Sunken Ships

My life as an ocean scientist was about more than expeditions. In my privileged career of four decades, I saw the small and the large: obscure creatures that lived perfectly normal lives in the ocean depths, on the one hand; the massive forces that define continents on the other. These discoveries could not go unremarked: They led to poetry, parody, and world-sized maps. Neither could I forget the ships on which I sailed. Thus I devote a chapter to the fates of great ships.

CHAPTER 1

Volcanoes on Earth and Mars

During my undergraduate and early graduate years, I considered a career in comparative planetology. While that never happened, I did think and publish on a global and planetary scale a few times in my career.

In 1972, I published a paper in *Nature* magazine suggesting that massive volcanism was the cause for the global extinction catastrophe that killed off many species, including dinosaurs. (We now say non-avian dinosaurs.) I pointed to the Deccan Traps in India, a vast expanse of layers of flood basalts formed from lava outpouring, which seemed then to about match the time of dinosaur extinction. However, about six years later evidence was found that suggested an asteroid colliding with Earth was responsible, which led to the later discovery of the crater buried under Yucatan and the adjacent shelf.

Already in 1979, I realized that large impact was far more powerful than any volcanism. However, those scientists who disputed the impact or impact-only explanation continued for some years citing my 1972 paper. Now it's clear that the impact happened during the period of the Deccan volcanism, which spread out over time and had begun before the impact. It remains curious that two very rare (about one every 100 million years) events, one geogenic and one astrogenic, should happen about the same time. I had to write a science fiction book, *What Really Killed the Dinosaurs,* Amazon, 2020) to explain this apparent coincidence.

In the early 1970s, I also published two papers examining the heights and spacings of volcanoes in mid-plate settings. For example: the Cape Verde and Canary Islands volcanoes on the African (Nubian) tectonic plates. I compared volcano heights with those on Mars, which are not broken up into plates but appear to have a single very thick rigid shell. The Martian volcanoes are higher and more widely spaced, so my paper suggested that this was due to the difference in thickness of the lithosphere (solid outer part of these two planets).

This Dynamic Planet

In the early 1990s from the Naval Research Laboratory, I began to collaborate with geologists from the US Geological Survey (USGS) and the Smithsonian Institution, three different government entities. This was informal collaboration; no funds were requested or transferred. We worked on a global map to be called This Dynamic Planet with volcanoes, plate boundaries, earthquake epicenters, etc. No culture; e.g. political boundaries, cities etc.

The first author was always the late Smithsonian Institution volcanologist Tom Simkin. The first version of this map had appeared in 1989, prior to my involvement. I became the ocean member of the author team on the second and third editions. The 1994 Dynamic Planet map, printed by USGS, became the best-selling map in USGS history. Schools and universities posted them on classroom walls or easels. The third and best Dynamic Planet map was published and printed in 2006.

The paper map is not that much in demand today, but only because it can be viewed online and printed. The Smithsonian maintains the digital version up to date, for instance by showing the earthquake preceding the 2011 tsunami that destroyed the Fukushima nuclear power plant. Our map has not really become out of date, but zooming into a part of it on a smartphone or laptop is not the same as viewing the entire 48-inch-wide map. Every home on Earth with wall space should have a full-size copy.

We authors worked on this project as our regular funded duties permitted. A major focus was on explanation—in the choice of symbols, keys, and captions—in a way that would be understandable by non-geologists. Every attempt was made to avoid technical jargon. We wanted our map to be accessible to students, teachers, science writers, and anyone with curiosity. At the same time we did not want to dumb it down so much as to be useless to geoscientists. Working on this intra-government agency project was one of my most satisfying career experiences.

CHAPTER 3

My Proposal:
Ocean Floor Mapping

Having dinner one evening in 1999 in our Southern Maryland home, I was explaining how poorly we know the ocean floors, or at least 80 to 90 percent of the ocean floors. Then a bold thought came to mind: Why not begin an international, long-term project to map the world's ocean floors from beach to deep sea trench? Would it be feasible with current technology, if spread out over enough years? Might it provide an opportunity for peaceful collaboration, with data shared by all, and likely be more economically valuable to humankind than the space program? It seemed a good way to start the new 21st millennium, which was just around the corner.

This project idea—later called Global Ocean Mapping Project (GOMaP)—would lead to a one-day workshop held at the Casino Magic in Bay St. Louis, Mississippi. We got the best deal there, and it was close to the Naval Oceanographic Office. (I wonder if any of those in attendance played the slots during coffee breaks.) The Chief of Naval Research, Admiral Paul Gaffney, agreed to fund the conference. I kept the workshop to about 30 participants and, except for two participants from England, all were from the US. Among those on hand were folks from the petroleum industry, who might supply survey vessels at cheap rates when sitting in port awaiting contracts. Military and academic ocean scientists and bureaucrats were there, as well as astronomers knowledgeable about how much better the Moon, Mars, and even Venus have been mapped in terms of topography, even though 70 percent of Earth is covered by oceans.

The workshop was very successful in terms of validating feasibility of such a project, both technical and financial. A reasonable time frame would be 25 years. The cost would be less than the Apollo Project but the benefits greater. The biggest hurdle? Political will.

A quarter century has passed, but the Global Ocean Mapping Project has yet to happen, or even to begin. Our several publications are read and cited on occasion. Ironically, interest in returning to the Moon and even sending astronauts to Mars has now been revived after decades of dormancy. Mapping solar-system moons with orbiters is ongoing.

Perhaps it was a bad omen when Casino Magic, a floating, boat-like structure, was destroyed by Hurricane Katrina in 2004. Indeed, it was ripped from its moorings and deposited next to Interstate I-90.

Chapter 4

Intruders Into Stuffy Science: From Tubeworm Poetry to Graphic Solutions to Problems of Plumacy

Many scientists—especially geologists but not so much engineers—have a great sense of humor. They crack jokes in lectures and tell funny stories on geology field trips. Cartoons appear in illustrated lectures. But what about research journals? Generally not.

A paper published in the prestigious journal *Science* reported on seismic wave (sound and shear) speed measurements on Apollo lunar rocks. The authors added a footnote stating they had also tested green cheese and found it resembled lunar rocks. There was outrage in some quarters. Joking had no place in a science journal. A bad precedent, it was alleged, that would undermine public trust in science and public funding. Perhaps.

In the 1970s, I got to know fellow geologist Bob Dietz, who was known to be debating creationists, particularly those who espouse young-Earth creationism and believe Earth is only 6,000 years old. Dietz and Jack Holden, an ex-paleontologist turned cartoonist, published a cartoon spoof of creationism. I got to know Holden by mail, and the two of us turned out a parody of the new geological concept of hotspots or mantle plumes, postulated to account for Hawaii, the Galapagos, Iceland, and similar locations of anomalous volcanism not explained just by plate tectonics. (Convection plumes are narrow columns of rising fluids, as in local thunderstorms.) I was spoofing my own field of research; the ideas for the cartoons were partly mine and partly his, but the cartoons were only by him.

I had no hope of getting our hotspot parody paper published in a major formal research journal, so I didn't even try. But some science journals are less formal than others. For example: *EOS*, published by the American Geophysical Union; *GSA Today*, published by the Geological Society of America; and *Geotimes*, published by the Geological Institute. I first tried one of those geological journals, but got a stern rejection. However Fred Spilhaus, who had been on RV *Chain* in 1964 and was now *EOS* editor, agreed to publish our Holden and Vogt cartoon paper in a 1977 issue. It was titled "Graphic Solutions to Problems of Plumacy." We invented the spoof term plumacy to imply addition to the concept of deep mantle convection plumes. *EOS* had never published papers like that, so maybe Fred got some flak for publishing ours. Yet it was obviously a hit. When I visited universities to give talks at seminars, I noticed many of our cartoons taped to grad student office doors. Thanks to Durham University Professor Gillian Foulger, our 1977 paper was reprinted with another: "Plumacy Reprise: Plumatic Asylum, Omak, Washington."

I collaborated with Holden several other times. In the 1970s, we worked on a cartoon spoof of the various diverse theories to explain the dinosaur extinctions. This was before the asteroid impact was discovered. Alas, this was never published, but *EOS* (1979) did publish a poll of its readers regarding which theories were preferred.

In 1979–'80 and 2005–'6, I commissioned Holden to draw some cartoons on issues relating to Calvert County, Maryland. He never visited here so I just sent concept ideas as sketches: Patuxent River pollution, highway congestion, suburban sprawl, and, most recently, a proposed third reactor at the Calvert Cliffs Nuclear Power Plant. Two local papers published all our cartoons in various issues. Alas, those papers are now gone, but the reactor was not built. River pollution has been reduced, but suburban sprawl and highway congestion are worse than ever.

In the mid-1990s, we at the Naval Research Lab and our collaborators discovered and sampled the Haakon Mosby mud volcano in the Norwegian Sea, which I write about elsewhere in this book. After the 1996 cruise we hosted a party for ocean floor scientists, and among those on the guest list was Sherrod Sturrock, a former director of the Calvert Marine Museum. After overhearing scientists in a serious discussion of tiny tubeworms recovered from that mud, Sherrod asked: How could anyone get so interested in seafloor worms? Then she turned to me and announced that she felt a poem coming on.

Sherrod indeed wrote a poem about a mud volcano and its tubeworms. Maybe the only one ever written on that subject. And it would be translated into the other three languages spoken by participating researchers: German and Russian, with the Norwegian participant also doing so for his tongue. Publishing the poem in a research journal was no big problem—because I was co-editor of this 1998 special volume of *Geomarine Letters*. The journal senior editor had left this decision up to me.

Leagues beneath the grey-green surface

rumbles Neptune's seething mound.

Oozing molten muck and seeping gas it reigns,

dark and forbidding, unbidden.

Its pressured rage spewing forth a vaporous plume into

the unsuspecting world above.

Its summit wears a frozen crystal lattice crown; White,

ice-like, unearthly.

Ice and fire entombed by Earth and water.

A towering monument encircled by ghostly moa

it reeks of primal violence, fractured crusts;

The stuff of legend gurgling from its core—

The heat of forgotten eons rising to the ocean floor. And

on its surface roils the wormed monster, guarding the

sacred fire from prying eyes.

Like twisted roots the vermicular horde writhe across

the mud;

White, worm-like, unearthly.

Life forms sucking the residue of decay.

—Sherrod Sturrock's English original

Kak pod morem-okeanom Chudnyj holm na dne morskom,

I poros on ne bur'yanom,

A nevedomym chervem.

Nad holmom metan kuritsya, Sozrevaet v nem gidrat,

A nad nim podo bno ptice Proplyvaet redkij skat.

Glubzhe v nedrah, pod koroyu Lava bul'kaet, pyhtya;

To kak zver' ona zavoet,

To zaplachet kak ditya.

To v razlom zamyslovatyj

V drug poddast ona parka,

Tak chto sam Neptun bradatyj, Ryavknuv groznoe

　　"Bez daty!", S voem plyashet trepaka.

　　　　　—L. Polyak's Russian translation

CHAPTER 5

Ships of Yesterday

Most of the vessels on which I researched ocean-floor geology were scrapped, otherwise known as broken down, years ago. This also is the fate of passenger ships, nowadays mainly used for tourist cruises. The typical lifetime of such a ship is around 30 years. However, ships sunk in wars survive much, much longer as seafloor wrecks. During World Wars I and II, somewhere in the order of 10,000 ships ended up on the ocean floors, often merchant ships sunk by U- boats.

Other ships have been lost in storms, run aground on reefs and rocks, or rarely and famously by collisions with icebergs. Think *Titanic*. Even when located, salvaging them from the deep ocean floor is generally not worth the cost. Of course, sunken Spanish galleons laden with gold are another matter. Seafloor shipwrecks have marine biological research value in their progressive colonization by diverse benthic organisms and serve as deep-sea habitats and ecosystems.

Some ships are sold and resold, repeatedly renamed, and variously reconfigured and repurposed. This has been the case for some of the research ships I've known.

However, none of those renamings can compare with the cruise liner Cunard *Adventurer*, on whose 1973 eclipse cruise Voyage to Darkness I told passengers about plate tectonics and chatted with Apollo astronauts and author Arthur C. Clarke. The *Adventurer* was sold to the Norwegian Cruise Line in 1977 and the liner renamed *Sunward II*. She cruised under that name in the Caribbean until purchased by a Greek operator in 1991 for Mediterranean and Aegean deployments, renamed *Triton*. In 1997, she was again resold, this time to Royal Olympic Cruise Line, which in 2005 sold the ship to a Maltese operator, Louis Cruise Lines. Under its final name, *Coral*, the cruise ship my wife and I knew as Cunard *Adventurer* was retired after 42 years of service and broken down in Alang, India, in early 2014. Other sites of ship dismemberment are or have been in Turkey and China.

Many ships have been lost over the years to storms, wars, piracy, collisions, or run aground on reefs. With few exceptions, research ships, a tiny fraction of all ocean-going ships, survived until being recycled into scrap metal. Often, ships of war were sunk at sea and never rediscovered. Yet. Famous warships either sank directly like the WWII vessel HMS *Hood*, or were scuttled by their crews, like the battleship *Bismarck*, later to be rediscovered and explored. The Royal Navy lost about 500 ships in WWI, the Japanese 334. The most famous and most investigated shipwreck is the 883-foot-long *Titanic*, its starboard hull ripped open by a large iceberg late on an April evening in 1912.

Some two and a half hours later, the *Titanic* sank 12,500 feet to the ocean floor. More than 1,500 people died (estimates differ) by drowning or hypothermia. The crew and third-class passengers were disproportionately impacted. All told, 710 people were rescued. The wreck, rediscovered in 1985, rests (and rusts) in two large parts, about a half mile apart.

Three years later, in May, 1915, during WWI, the Cunard liner RMS *Lusitania* was torpedoed by the German submarine U-20 11 miles south of the southern Ireland coast. Of 1,959 on board, 1,195 perished. The *Lusitania* sank in only 18 minutes. Besides passengers and crew, the vessel carried US munitions from the still officially neutral US to the United Kingdom. The seafloor wreck of the *Lusitania* also has been explored.

In terms of lives lost, the sinking of the German cruise liner and wartime transport ship *Wilhelm Gustloff* in the Baltic on January 30, 1945 was by far the worst maritime disaster in global history. Crammed with refugees fleeing the advancing Soviet army, the vessel was torpedoed by Soviet submarine S-3 and sank within about one hour. Some 9,343 people (plus or minus), more than half children, perished. Many died—as many on the *Titanic*—of hypothermia in the freezing water. About 1,239 passengers were rescued. The Polish government declared the wreck a war grave and prohibits diving on it.

Some ships and subs became war prizes before being scrapped. A very few are preserved in sunken states as war memorials. The USS *Arizona*, the *Missouri*, *Utah* and *Oklahoma*, all sunk by Japanese bombs at Pearl Harbor, are national memorials or monuments.

When resources were available, vessels of historic significance were preserved and are open to the public, as is the case with the sloop-of-war USS *Constellation* in Baltimore's inner harbor. The *Constellation* was the last sail-only warship built for the US Navy, in 1854. Even older, dating to 1797, and more famous, is the battleship USS *Constitution* in Boston, the oldest surviving warship afloat in the world today.

One of three surviving Liberty Ships is tied up in Boston for public visits. The other two survive in San Francisco and in Greece. Largely serving as troop transports during WWII, more than 2,700 were built, the largest number of vessels in one class in global history. Almost all Liberty Ships were eventually scrapped. However, about 200 were sunk, some by U-boats and a few others by storms. In 1970, our vessel *Lynch* found Liberty ship *Le Baron Russell Briggs*, which remains on the ocean floor as a historic reminder of time when surplus vessels were used for ocean hazardous waste disposal.

Four German WWII U-boats of about 1,200 built survive intact. Some 800 were sunk by Allies, their remains resting (and rusting) on the ocean floor. Far fewer U-boats were built in WWI, and of those, more than half were lost. Late in WWII, Germany's technologically advanced U-1105, *Black Panther*, surrendered to the Royal Navy days before the Nazi surrender. The sub crossed the Atlantic in January 1946 and was given to the US Navy. As commonly with captured ships of war, the U-Boat was used for target practice in an operation known as Sinkex, nowadays carried out only once a ship is towed at least 60 miles to sea. Sunk and retrieved five times, U-1105 rests in 91 feet water off Piney Point in the Potomac River, where it remains as a National Oceanic and Atmospheric Administration Historic Shipwreck Preserve, popular among amateur divers not deterred by turbid water and strong tidal currents.

In 1982, the USSR scuttled a very contaminated (radioactive, due to reactor failure) nuclear submarine, K-27, in the Kara Sea—even though disposal of high-level radioactive waste was banned by the International Atomic Energy Agency in the 1950s. However, a few nuclear submarines have been sunk by accident, among them Soviet submarine K-278 *Komsomolets* (lost with 43 fatalities but 27 survivors, due to engine room fire, April 7, 1989) revisited and investigated by Russian nuclear experts and one geoscientist via Mir dives on our 1998 *Keldysh* expedition.

All told, the Soviet Union lost five nuclear subs and its successor, the Russian Federation, lost two more. The second and latest K-141, *Kursk,* sank in the Barents Sea on August 12, 2000, due to explosion in its torpedo hull. By contrast, the US only lost two nuclear subs (*Thresher* on 10 April, 1963, and *Scorpion*, 22 May, 1968). There is little doubt that US submarine construction, maintenance and operation is more careful than in the Soviet and Russian navies. My limited experience on Soviet and Russian research ships suggests that this difference in attention to safety is pervasive.

The USSR loss in March, 1968, of K-109 is a special case because the United States, not the Soviets, located the wreck northwest of Hawaii and secretly tried to recover at least part of it in 1974. The CIA got Congress to fork over $800 million to design and build the *Glomar Explorer*. To cover up mission goals, the CIA announced that *Glomar Explorer* was looking for manganese nodules, a potential source of various valuable metal oxides. A small portion of the wreck and the bodies of several Soviet sailors were recovered.

The timing of this operation and the vessel name caused public confusion with the research ship *Glomar Challenger*, on which I served as co-chief scientist in summer, 1975. Both vessels were built by the Howard Hughes-connected Global Marine Corporation. I know some folks believed I was somehow involved in recovering that Soviet sub and its Cold War secrets. The *Glomar Explorer* was scrapped in 2015.

The ultimate fate of the many iron-hulled shipwrecks resting on the global ocean floors is to rust away very slowly, provided the water has enough dissolved oxygen. Repeated exploration of the *Titanic* wreck is tracking the process. Wooden hulls disappear more rapidly, consumed by many types of organisms. However, shipwrecks on the anoxic bottom of the Black Sea or deep Norwegian fjords may last for millennia.

No research surface vessels have been preserved for history, to my knowledge, and there have been only a very few post-1900 losses. However, shipboard injuries and fatalities have occurred. The nonmagnetic ship *Carnegie*, built in 1909, was lost to fire and explosion in 1929. A Japanese research ship was destroyed in 1952 with all 31 on board killed when working above the active sub-marine volcano Myojinsho. In 1987, the research ship RV *Melville* was shaken by the "fearful clamor of large steam bubbles bursting under its hull" when the sub-marine volcano Macdonald Seamount erupted in the Pacific Ocean only 130 feet below the vessel. In my ocean geology career, I have published several technical papers on sub-marine volcanism but was never on board a vessel investigating one of the few active or dormant ones. However, I likely would not have turned down the chance unless an eruption was in progress.

No manned research submersibles were lost until *Titan*— never properly tested—imploded on June 18, 2023, with all five on board instantly killed. The implosion has been attributed to failure of its carbon-fiber pressure hull while on a tourist dive to the *Titanic* wreck. In 1968, the Woods Hole submersible *Alvin* sank to the 1,500-meter-deep seafloor when the cable used to raise it snapped. No one was inside. It was recovered a year later, the onboard lunches surprisingly well preserved by the high pressure and low temperature. Alvin has been repeatedly upgraded, its current dive limit being 21,335 feet. Only a few deep sea trenches are deeper.

Mir-1 and Mir-2 were decommissioned in 2017 and are presently on museum display in Kaliningrad. However, *Akademik Mstislav Keldysh*, of *Titanic* movie fame, is still active in research. NR-1, sometimes nicknamed Nerwin, was decommissioned in 2008 and scrapped. However, the NR-1 control room was preserved and in 2018 installed in the US Naval Undersea Museum.

Already in the 1970s, Admiral Hyman Rickover, the father of NR-1, was planning for an NR-2, but Congress was unwilling to fund the project. Rickover's influence had waned. So there would be no NR-2. In recent decades, the technology of unmanned robot seafloor mapping vehicles—autonomous underwater vehicles—had advanced so far that manned subs like NR-1 and submersibles like the Mirs are largely not worth the cost. Similarly, the robot research of, say, Mars and the Moon is far more cheaply done by robot landers and orbiters.

In summary, of the 18 ocean-going vessels of my career (17 research ships and one a cruise liner on which I lectured), at least six remained in service in 2024. *Healy* and *Cory Chouest* (United States); *Marion Dufresne* (France); *Haakon Mosby* (today ACC *Mosby*) (Norway); *Kurchatov*, *Logachev*, and *Keldysh* (Russia). I don't know what happened to the *Mendeleev*, but we never departed Kiel.

Haakon Mosby was converted to a fisheries research trawler and renamed. *Cory Chouest*, NR-1's last mothership, has stayed in service for special Navy missions, including surveillance. *Keldysh* remains as Russia's largest research vessel. *Healy* and *Marion Dufresne* operate today as they did when I was onboard.

PART VI

Home Base: Chesapeake Bay

It might seem unlikely to you, reader—as it must have many times to my wife Randi and sons Erik and Jason—but I had a home life as well as my work-a-day life as a far voyager. Chesapeake Bay, my home for seven decades, had its own history written in its depths. I had the good luck to help recruit ships both small and large to delve its bottoms for those secrets.

From my home base, I helped unravel stories of present and past Chesapeake Bays, taking sediment cores aboard the *Marion Dufresne* (1999, 2002) and doing dayboat geophysics, the last on *Rachel Carson* in 2010.

Surprise gifts of ice in extraordinarily cold winters gave me new acquaintance with old-time pleasures. I even helped a team of scientists test the effluent, in the form of 'poopcicles' of Bayside homes. I end my Chesapeake narrative with a personal history.

CHAPTER 1

Getting to the Bottom of Chesapeake Bay

For my first 25 years living along Chesapeake Bay and seeing it from our house looking down a ravine, from a nearby beach, or from a small sloop-rigged dinghy, I envied geoscientists who studied what is on and below the bottom of the Bay, yet slept at home on most nights. My research required a month or more away from home on the deep oceans or in aircraft.

Finally, in 1994, I got the chance. Our division at the Naval Research Laboratory had begun to research shallow water venues. Maybe this was a switch in priorities away from finding Soviet (now Russian) subs and hiding ours. Colleagues had developed a geoacoustic system, using a transducer-studded towfish that estimated the physical properties of bottom sediments. And they wanted to test it in the Chesapeake Bay.

No, the United States was not expecting Iran or North Korea to sail up the Chesapeake and land troops. The idea was to test the new system in the Bay, not in a hostile region like the Persian Gulf.

Why? Ground troops need to know about bottom sediments when they land. Systems like we tested might detect buried mines and even *proud* mines (those partly or entirely above the bottom).

Meanwhile, our DC branch acquired an Edgetech system that included a side-scan sonar (100 and 500 kilohertz, far above human hearing) and a sub-bottom profiler, called

Chirp—because the sound emitted is not a bang or ping but a sweep of frequencies (1–12 kilohertz, audible to human ears). This helps computers distinguish echoes from unwanted noise.

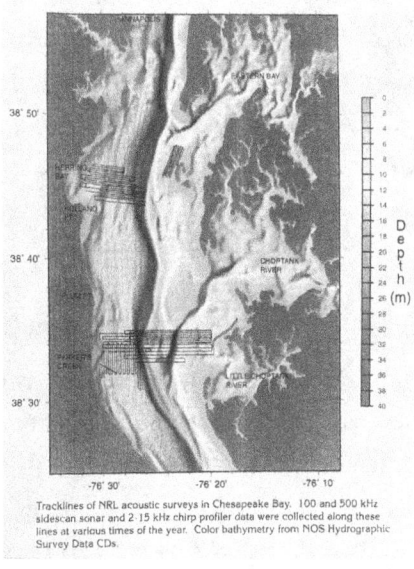

Tracklines of NRL acoustic surveys in Chesapeake Bay. 100 and 500 kHz sidescan sonar and 2-15 kHz chirp profiler data were collected along these lines at various times of the year. Color bathymetry from NOS Hydrographic Survey Data CDs.

Acoustic surveys from our vessels revealed
what was on and below the Bay's bottom.

With this profiler, we could get echoes back from as deep as 100 to 200 feet below the bottom of the Bay. Of course the echo from the bottom always comes first, loud and clear.

We compared the two systems by repeating some tracks. I finally was involved in doing science in Chesapeake Bay. Our lab did not have the proper boat for this research, so we hired the Maryland Geological Survey's RV *Kerhin*, a multipurpose vessel based at Sandy Point State Park just north of the bridge to the Eastern Shore. I spent many days aboard from 1994 to 1999 and later.

Kerhin was first called *Discovery* but renamed after the assistant director of the Maryland Geological Survey, Randal Kerhin, who died unexpectedly in January 1999.

The vessel was 51 feet long with a 16-foot beam and had a small galley with a fixed table and two benches. Up to six could sit there. Nearby was a sink, a refrigerator, and a microwave. The skipper back then was Rick Younger, whose dad had also captained Bay research boats. Rick's crew numbered one. While it was much smaller than the deep-sea vessels I was used to, *Kerhin* could zip over the water at 20 knots and up to 24, twice as fast as any of the blue-water ships.

I call boats like *Kerhin* day boats because one can get out of bed, spend the day on the water, and be back home that night. What luxury! Sometimes we parked our cars at two places and debarked elsewhere. In Solomons, at the bottom of Calvert County, for instance.

Technically, one could bunk on one of those galley benches, but I never did. As in diving on Russian Mirs in 1998, it's a very good idea not to poop onboard *Kerhin*, and God forbid you have diarrhea. It's possible to get sea sick due to small vessel size, but looking out at the adjacent Bay is easy and helps.

The present Chesapeake Bay (yes, it had predecessors) began to form around 10,000 years ago, as rising sea levels first pushed Atlantic tides into the Susquehanna River valley north past the Virginia capes. That's Cape Charles and Cape Henry, where Norfolk is today.

The melting of global ice sheets—especially the great Laurentide ice sheet that covered most of Canada—had advanced south into Pennsylvania, about to the location of Scranton and Wilkes-Barre today. At that time, 20,000–

25,000 years ago, the Atlantic shoreline was near the edge of the continental shelf, 50 miles east of the Virginia capes. The present Chesapeake Bay area was covered by boreal conifer forests and patchy tundra. There was even permafrost in the northern Chesapeake area, along with mammoths, mastodons, saber tooth cats, giant ground sloths, and short-faced bears.

Native American tribes of the Early Archaic period were living here when the Bay first began to form, still narrow and partly freshwater tidal, like the Patuxent River still is today. As sea levels continued to rise 9,000–8,000 years ago, tidewaters began to flood low plains and penetrate up into tributary valleys, like the now-submerged and buried Ice Age part of Parker's Creek, in Calvert County, Maryland.

NRL's Mike Czarnecki works on towfish
that sends out acoustic pulses and records echoes.

As the Bay formed, sediments began to accumulate on the bottom. They were a mixture of materials eroded from adjacent lands plus dead organic matter, mainly from planktonic microbes but also windblown pollen from surrounding forests. This occurred season by season, year by year. Although disturbed and mixed by benthic organisms, these sediments provide a record of what lived and died in and around our growing estuary.

This priceless record is being read by modern science in the form of tubular cores extracted from the soft mud. Because the sediments below the modern Bay are still soft, they can be cored by dropping a steel pipe into the bottom. Inside the pipe is a smaller PVC pipe that is pulled out of the metal core barrel once the pipe is back on deck. A crown-shaped core catcher is always inserted into the bottom of the pipe before the core barrel is lowered by cable. This keeps the soft sediment from sliding back out into the water as it is being pulled up.

The PVC pipes are sliced along their length and one half preserved as an archive. The other half is sampled by specialists. An age-versus-core depth below the Bay floor table is constructed using Carbon 14 dating on wood fragments, seeds, and other sediment elements. Colonial settlement and land clearing created a proliferation of the Ambrosia ragweed pollen, recorded by pollen blown out over the Bay that created a marker age in sediment cores.

Palynologists using microscopes sort out pollen by species or at least genus so forests growing in our region in the last 10,000 years can be reconstructed. Because plankton species change with salinity, wetter and drier periods were recorded. Watershed droughts mean less river input and more Atlantic saltwater.

Sediment coring and core analysis cost time and money, so coring sites are carefully chosen, with the help of sub-bottom Chirp profilers and side-scan sonar mapping. Our Naval Research Laboratory work on RV *Kerhin* in 1994 and 1999 provided road maps on where to core and where not to.

A few Bay bottom areas have been swept clear of young, soft sediments due to currents. Below, the soft bay-floor sediments are older, more compact sediments like those exposed in the Calvert and Nomini Cliffs. Such sediments have to be cored with a rotary drill. Dropping a steel pipe core barrel onto such sediments usually serves only to bend the barrel and turn it into scrap metal.

Sands are hard to penetrate and often fall out of a core barrel as it is pulled up. A bigger problem is tiny methane bubbles in sub-bottom sediments. Methane (natural gas) is produced as waste by bacteria in their consumption of buried organic matter. It's really much as humans and other animals produce methane from digestion of food.

Even a very small percentage of methane bubbles reflects or scatters sound waves and thus blocks seismic energy from penetrating into and reflecting from sub-bottom layers. As recently as the early 1970s, some geologists thought there was hard rock below the bottom.

Below the central channel of the Bay the sediments are so gassy that there is no choice but to core blind.

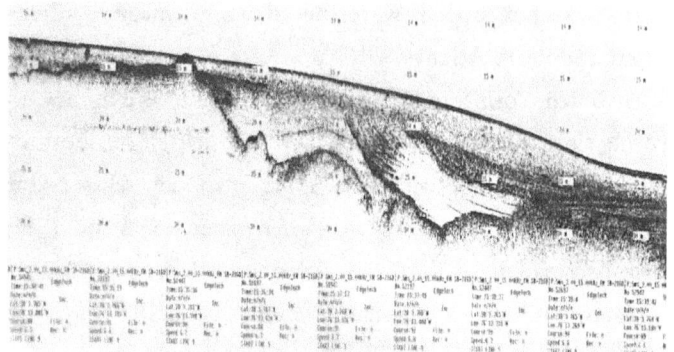

A profile of bottom-water depth in the Chesapeake Bay
with the echoes of an older, rougher bottom that got buried.

The smallest research vessel in my long career was the RV *Kerhin*, on which I took many day trips using sub-bottom Chirp profiling and side-scan sonar mapping. The *Kerhin* was the perfect size for the Chesapeake at 51 feet long with a 16-foot beam. It seems very incongruous, but I also researched below the Bay aboard the second largest vessel of my career and probably the biggest ship ever to conduct Chesapeake research.

That happened in 1999 and again in 2002.

It was the French ship *Marion Dufresne*, the surname of a colorful 18th century French explorer. Built in 1995, the *Marion Dufresne* is one of the world's larger research vessels at 395 feet long with a 28-foot beam, weighing 10,400 tons. The vessel had 10 officers, 20 crew and room for 110 passengers.

The large size reflects its additional non-research functions: resupplying the French islands such as Kerguelen and Crozet and the French Antarctic bases. The ship is also available for rescue missions. The stormy, 25- to 50-foot seas of the Roaring Forties and Furious Fifties required a large ship. Winds there exceed 60 miles-per-hour 100 days per year.

Occasionally the *Marion Dufresne* even carries a few paying passengers, people with strong stomachs willing to spend up to 60 days at sea between ports.

What brought this ship into the Chesapeake Bay? The short answer is its unique CALYPSO sediment coring system, which is capable of cores up to 190 feet long if sediments are reasonably soft. Much core analysis is done onboard in what amounts to a floating laboratory. Sediment not immediately sampled or analyzed is refrigerated for later cold storage. Refrigeration is essential to prevent further bacterial consumption of organic matter.

The *Marion Dufresne* regularly visited the French Caribbean islands and from there traveled along the Atlantic margin to French Canadian Quebec. My US Geological Survey colleague, Tom Cronin, had the brilliant idea of hiring the vessel to make a slight detour, going north via the Chesapeake Bay and through the canal and Delaware Bay, then back into the offshore Atlantic.

Until 1999, the large number of sediment cores, almost all less than 15 meters long, had only sampled the last few thousand years of Chesapeake history. The Ice Age channel-bottom of the lower Susquehanna downstream of the present river mouth is around 180 feet below present sea level, as estimated from seismic reflection surveys from 1988 to 1990. Actual core sampling and channel depths were known only from engineering drilling prior to construction of the Annapolis Bay Bridge and the Chesapeake Bay Bridge-Tunnel.

The present ship channel is what remains not yet fully buried of the Ice Age lower Susquehanna Channel. In the far north and far south, the old channel was fully buried and is kept open by dredging. Along the rest of the channel, sediments vary from about 30 feet to 165 feet thick, but average around 100 feet. Only the *Marion Dufresne* could retrieve 100-feet long sediment cores with their up to 9,000 to 10,000 years of Chesapeake paleoenvironment.

I was involved as junior co-chief scientist, planning the Naval Research Lab site surveys, interpreting those data, helping choose the core locations, and being on board. In 1999, I was onboard only for a few days; had to get off via pilot boat before the last coring because I had a date with Submarine NR-1 in Bergen, Norway. I spent more time on the *Marion Dufresne* in 2002, flying to Miami to board.

I suspect it was bad luck to spend the least research ship time on the most comfortable vessel. The French approach was to keep the shipboard experience as civilized as back home. The galley served meals on white tablecloths and with wine. For those not on watch there was even a small dance floor with taped music.

Both 1999 and 2002 *Marion Dufresne* corings were very successful. Several cores penetrated to the Ice Age river-channel bottom. The longest penetrations were near the mouth of the Potomac estuary (26 meters penetration and 20.7 meters sediment recovered) and west of Kent Island (23 meters penetration and 17.2 meters sediment). This depth exceeds all but one previous core. The older sediments in all the long cores indicate freshwater tidal environments similar to those in the modern Potomac from well south of Mt. Vernon into Washington DC.

Up to 1999, the *Marion Dufresne* had never cored in water depths less than 400 feet. But I proposed a core into the paleochannel of Parkers Creek in Calvert County, Maryland. The channel bottom was not obscured by methane near its margin. This site was less than 1.5 miles from the Calvert County shore in only 35 feet of water. This was only 10 feet below the vessel's draft.

The ship's senior engineer was a bit flummoxed. The standard scheme was lowering the core assembly on a long cable, then triggering its final plunge into the bottom. The savvy operations chief, Yvon Balut, also was at first flummoxed, then decided it could be done.

The *Marion Dufresne* was equipped with a Calypso core-retrieval assembly, then the best in the world by virtue of the deepest penetration by the core barrel and the longest cores retrieved. Balut had his Calypso team strap the cable alongside the hull on the starboard side below the railings. The straps would come off as the core assembly was dropped into the Bay from the crane. Everyone then cleared the deck. In a seeming instant after it was released, the 10-meter-long core barrel fell into the water. All 10 meters cored into the sediment; 7.82 meters of sediment were recovered.

The core barrel was stopped by ancient oyster shells, which meant it probably reached to the depth of the paleochannel. An oyster shell was later dated to an age about 8,500 years. This meant that rising sea levels had already flooded the Parkers Creek Ice Age valley to within 1.5 miles of the modern creek mouth, at sea levels still nearly 18 meters below present sea levels. And we now know, the young Bay already supported oysters 8,500 years ago. I wonder if local natives were gathering and eating them. It would take another millennium before rising sea levels overtopped the Parkers valley and began forming and eroding the modern Calvert Cliffs.

That core was taken on Tuesday, June 22, 1999 at 6:30 a.m.

The giant blue ship sat at anchor for several hours not far out. It should have been conspicuous and even alarming as seen from Scientists Cliffs and Dares Beach. Yet I never met anyone who saw it. Maybe most people were still asleep or already on Maryland Rt. 4, commuting to Washington DC.

I later served on the Bay's own research ship, the Canadian-built *Rachel Carlson* owned by the Chesapeake Biological Lab in Solomons. I was co-chief-geophysics/acoustics scientist on an early cruise in the 20-teens led by Dr Lora Lapham, who has been studying biogenic methane in Bay sediments. Tom Miller, then Chesapeake Biological Lab director, subsidized this cruise in the Chesapeake mid-Bay to help promote research applications.

Rachel Carlson was, of course, the marine biologist credited as the founder of scientific environmentalism. Her classic book *Silent Spring*, focused on agricultural pesticides, especially DDT, and their effect on bird-egg shells. She lived in Baltimore, studied at Johns Hopkins, taught at the University of Maryland and worked most of her life for the US Fish and Wildlife Service. Her work was celebrated by environmentalists but made enemies in the Ag industry.

At the vessel arrival-in-Solomons ceremony, a smartly uniformed NOAA official raised her hand and asked "Who is Rachel Carlson?" Memories are short.

The fine vessel is underutilized due to high operating cost, despite being faster which reduces transit costs. In today's science funding climate, the *Rachel Carson* is likely to serve mostly as a landmark for yachts entering and exiting the harbor.

CHAPTER 2

Winter Sports in Southern Maryland

Southern Maryland is hardly a winter sports paradise. Not even close. But in that occasional winter, Canadian cold arrives. A cold air mass has to blow its way south fast; otherwise it warms. Traveling over snowy ground helps keep the air mass cold because snow reflects the warm sunlight and radiates out to space any warmth on clear nights. A few successive days with lows in the mid-or low teens and highs in the 20s is enough to turn our small (maybe one acre) community pond into a skating rink.

The winter of 1976–'77 provided marvelous days of skating.

Alas, kids can't resist throwing out rocks and sticks, which freeze into the ice surface and trip skaters. Still, I recall many happy days skating on our pond in the 1970s and 1980s. We played backyard ice hockey and used straight branches to create goals. One cold night we even laid out Mexican-type candles in sand-filled paper sacks to make skating paths on the pond. Some winters we built warming fires on the pond's shore. Knowing more about ice than others, I was the ice checker, deciding when the ice was safely thick. I made sure there was always a long bamboo pole—just in case someone broke through and needed help getting out.

One day while walking out on the ice to check thickness, I thought I saw, through the ice, a large, round moss-covered rock on the bottom. Being a geologist, I realized that was unusual in Southern Maryland. Dropping down onto the ice for a closer look, I was shocked to see this rock, only a foot below me, slowly moving. It was a snapping turtle! I learned that they can hold their breath for a long time while absorbing oxygen dissolved in the water.

Occasionally a winter would bring major snow, which generally melted away in a few days. Some of us who had skis skied on our surrounding hilly woods. Lacking cross-country skis, I unlatched part of the bindings on my giant slalom skis. That way ski boot heels were free to pivot above the skis.

The winter of 1976–'77 was likely a once-in-a-century cold winter. It stayed cold from fall through much of the winter. The strong, persistent northwest winds kept blowing new Bay ice forming off our shore to the Eastern Shore, where many fishing boats were frozen in for months. Very little snow fell. This unusual situation caused most of the tidal Patuxent to freeze.

I commuted on Route 4 daily across that river on the bridge near Waysons Corner. One January day, I stopped there and checked the ice. Clear and thick enough to walk on. This would likely never happen again in my lifetime. So I found a Naval Research Laboratory colleague, Skip Kovacs, who said he owned skates and would go along with me skating 20 miles from that bridge to Lower Marlboro in Calvert County. That stretch of river was tidal, but there was little or no salinity to weaken the ice.

We packed spare clothes and a long rope and brought along a sled to be towed behind us. Our cars were parked at both ends. I brought along my 8mm camera and documented this rare experience. We had the entire scenic river to ourselves, barring only a few Canada geese who left their poop on the ice in a few spots. The only bit of open water was near Nottingham, and we never skated near it. We had the northwest wind behind us and could open our jackets as sails. Only where the Patuxent meanders back towards the northwest did we have to skate against the wind. I will remember that special skate to my dying day. Now where is that 8mm movie reel?

The Chesapeake Bay is not only an ever-changing canvas but also a blackboard. Written on this board are lessons in fluid dynamics and optics. And in those occasional cold winters also lessons in ice. The Bay off Southern Maryland is brackish, with salinities averaging around 10 parts per thousand, about a third of open-ocean salt. Bay ice, like the sea ice I experienced on icebreakers, differs more than a little from fresh-water ice. In fact, the first stages of winter sea ice formation in the Arctic and Antarctic can be observed on our

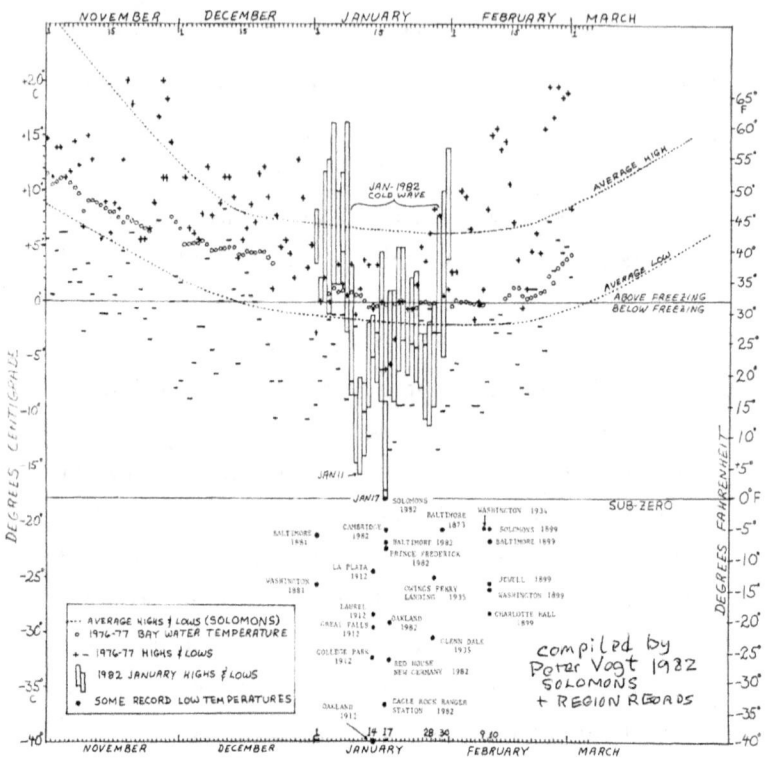

In 1982, I put together a chart plotting winter highs and lows over time in the mid-Chesapeake Bay region. The first sea ice is tiny short needles of relatively fresher ice as freezing salty water tends to exclude salt. Frazil ice, as it's called, gives the ocean (and brackish Chesapeake) a greasy appearance, as the short water ripples are damped out. On my 1965 and 1966 icebreaker expeditions, we remained up in the Barents Sea long enough to see frazil ice in the September beginning of the annual freeze up.

Chesapeake waves with frazil ice break gently on a beach, with soft whooshy sounds, not splashy ones. Frazil ice needles may be washed on the beach by onshore winds. The accumulated needle masses form slushy ridges that freeze together into hard white crusts. These crusty strips, which are not at all slippery, make great clean walking paths. You don't sink in, as on sand. One winter, with the eager participation of neighborhood kids, we built a kind of igloo, using a spade to cut frozen frazil ice into building blocks. We roofed over the structure with driftwood. Of course our igloo, perhaps unique in Bay history, was eliminated a few days later by Chesapeake waves and warmer weather.

One cold winter my 10-year-old son and I hiked along such a ridge of frozen slush for more than a mile, ice skates slung over our backs. We skated up a small but locally famous tidal creek, Parkers Creek, for more than a mile. The tide water moving into and out of the creek had raised and lowered the ice cover, creating a crack that ran along the creek middle, much as a median stripe along a county road. The creek was about that wide. Water had been squeezed out of the crack, flooding and paving the sides of the creek in new slick ice. We turned back when we encountered open flowing water.

Upon emerging from the creek we were surprised by—and surprised—a well-known Maryland politician, state Comptroller Louis Goldstein. As owner of the land on the north bank of the creek, Mr. Goldstein wondered if what we had done was safe. However he was also a hunter and had a duck blind along the beach nearby. In fact he had just bagged a duck, which had landed on the broken but apparently solid offshore ice. The comptroller offered the duck to us—if we could get to it. However, we respectfully declined.

If the air remains cold, frazil sea or brackish Bay ice forms small floes of slushy ice. It's soft because it's a fabric of frazil ice needles and saltier water pockets. As the floes bump into each other in the waves, they form round pancakes with raised rims. Indeed, this is called pancake ice. If cold persists, smaller pancakes freeze onto others, so ever bigger pancakes are formed. Eventually the ice becomes continuous, and once sufficiently thick can be walked on.

In 1977, it was cold enough to build igloos on the Chesapeake shore from washed-up frozen frazil ice.

If the air is calm, the nearshore Bay could freeze without breaking into floes. This happened just before the Washington's Birthday (today called Presidents Day) blizzard in February, 1979. The Bay had frozen over entirely except where ships had kept it open. Due in part to low tides and lack of waves, the water between two sandbars and the shore had frozen smooth. I carefully checked this ice for safety. Then the aforesaid son and I both went skating. It was and likely will remain the only time anyone in our community ever skated on the nearby Bay. The brackish ice felt rubbery, bending but not breaking under my weight. Afterward, I threw my hockey skates back into my VW Rabbit's trunk for my next skating-rink skate. The next day we got nearly three feet of snow. The following fortnight I went skating at a rink, a modest detour on my way back from work in DC. My skate blades were covered in rust; they had remembered their unique adventure on brackish ice.

The combination of waves and thick pancake ice can cause floes to ground on sandbars and can create an extra-terrestrial Bayscape. The shore becomes eerily quiet except for occasional faint cracking sounds. Floes are shoved against and over each other, forming jagged miniature mountain ranges two feet and even more in height, running along the crests of nearshore sandbars. Thawing creates small icebergs that float away and soon melt. A giant iceberg of that sort, perhaps 10 feet high, formed near the Eastern Shore at the end of that 1976–'77 Ice Age winter and made the DC and Baltimore news.

Thawing winds, waves, and tides soon clear brackish Chesapeake Bay beaches of all ice. A major freeze is forgotten.

Or is it? Hold on; not so fast. Thick, solid freshwater river ice remains in Chesapeake tributaries for days, so Bayside residents are surprised to see armadas of ice traveling south, carried by Coriolis Force tidal currents in narrow convoys along the Western Shore. (The Coriolis Force is the result of the Earth's rotation, which causes the counterclockwise circulation in the Chesapeake. The average current flows north along the Eastern Shore and south along the Western Shore.)

In polar seas, briefly digressing from the Chesapeake, the annual ice can be three to five feet thick. At polar air temperatures of minus 40 and below, even salty sea ice freezes solid, including tiny pockets of salt water. Most modern icebreakers just plow through the gray-green annual ice. However, over several seasons of summer melting, sea ice gradually becomes fresher. Repeated freezing and thawing excludes the salt. Old polar sea ice is not only thick but rock hard. Anyone riding on an icebreaker notices the difference between the swishing of annual ice sliding along the hull and the wham sound of hitting old polar ice. Old polar ice also is a beautiful powder blue, especially when the breaker turns a small thick floe upside down. Ice itself is, of course, not colored. It's all about scattering of sunlight, which is why the sky and oceans look blue.

CHAPTER 3

The 'Poopcicles' Saga

I never became an ice or snow researcher, so all the above remarks are anecdotes and natural history observations. However, a few years ago I had the chance to advance science in one small way that involved ice. Icicles, to be precise.

The Calvert Cliffs, bordering our community, rise 50 to 100 feet above the beach. The Miocene layers—comprised of marine silts, clays, and fine sands, plus or minus fossils—are laid down in a shallow, Atlantic embayment from about 15 million years ago at the base, to about 11 million years ago at the cliff top. Except for the oldest layers just above the present beach, these sediments are permeable to water: rainwater, gardening water, treated wastewaters, and waters leaking from septic systems of houses on the cliff. Water seeps down through the various upper layers and then seeps out the impermeable layers at the bottom, keeping these lower layers moist.

This water would be hard to collect and, for that reason, has never been analyzed. If all the septic systems along the cliff tops are working correctly, the seeping water should not be polluted.

During cold winters, the water seeping out of the lower cliff freezes and forms ice layers or icicles. One cold winter, I hosted a class of fourth graders so they could look at the layers with fossils and hunt shark teeth on the beach. I always warn visitors of all ages not to dig into or climb on cliffs. However some students, probably boys, decided it would be fun to lick the icicles. It had not occurred to me that anyone would want to do that. Were they thinking of popsicles? As far as I know, no one contracted any disease from septic-tank discharges.

Grad students had the task of gathering these rare ice formations.

Some years later I attended a lecture at the University of Maryland Chesapeake Bay Laboratory. A professor there and his PhD candidate were using conservative compounds to measure septic wastewater transport through sediments to the Chesapeake tidewater. Conservative compounds are those that neither humans nor microbes can digest, so they pass through our guts and through septic systems and become tracers, like dyes, of groundwater transport. Examples include artificial sweeteners, which can be measured even in tiny amounts using mass spectrometry.

The housing development where the PhD student was working lacked enough homeowners who were willing to allow their septic tanks to be sampled. Were the others hiding something? The possible sources of conservative compounds obviously required sampling. So we offered to let the student sample our tank, as did some of our neighbors. It was better than no tank samples. In the end, the candidate found enough willing homeowners in her test community.

Then it occurred to me that our own community's icicles could be sampled and tested for conservative compounds. I suggested this approach to the professor and his grad student, and it happened: Ice was collected during two cold snaps. I hauled out my Innsbruck ice ax to assist in the collection of the ice in places where simply snapping off icicles didn't suffice.

Some sites with ice were clean, while several others had dissolved sweeteners. Finding in the latter sites did not prove there was pollution. But neither was it ruled out. I presume the results found their way into the doctoral thesis. Yes, this was snooping into dietary habits of clifftop residents, and I did inform one of them. Perhaps this can be worked into some future whodunit with scientists deploying forensic biochemistry?

The idea is yours for the writing, dear reader.

Of course I was not so thrilled shoveling snow from sidewalks and scraping ice off car windshields. That happens occasionally even here in Southern Maryland, where I have spent most of my adult life.

There was only one time in my life when I truly hated ice. It happened one winter day about 1980 or 1981, right here in warm Southern Maryland. The thermometer actually dove below zero, maybe minus 5 or minus 8. The water pipes and insulation where I live were unprepared. Our plumbing had frozen; I thawed the pipe with a propane torch, keeping well away from anything flammable. Or so I thought. Just before heading off to DC for my distant day job, I got a faint whiff of smoke from another spot along the side of the house. Somehow I must have ignited some insulation.

I called the fire department and, using axes, they put out what was smoldering. The fireman said if I had gone to work, the smoldering would have continued for some time, then the house would have burned down. I was in shock. The fireman consoled me: "Don't feel bad. Plumbers burn houses down all the time."

These days I get extra sad learning of the ice sheet and glacier retreat from greenhouse gas-caused climate change.

CHAPTER 4

Vogts gate: Home with Heritage

This sign has hung next to our right front door along the Chesapeake Bay shore for decades. It appears to be a misspelled street sign, with no apostrophe and a missing capital.

Right and wrong. It was once a street sign in Oslo, Norway. But not misspelled; means Vogt Street. *Gate* means *street*. A cognate, the same linguistic derivation in English and Norwegian, both Germanic languages.

My home has an original identifier via Oslo.

Here is the story of this sign. Accompanied by family, I spent the school year 1977–'78 in Oslo, where I was visiting professor. Away from, but sponsored by the Naval Research Laboratory. But that's another story.

The Geology Department staff and grad students held a farewell party for me in early September, 1978. I got several wrapped gifts. One was a triangular moose-crossing-traffic sign, which they bought because I thought they looked cool. I had it here in Maryland for many years and eventually gave it to Autumn Phillips-Lewis, the land manager of the American Chestnut Land Trust, who said that she thought moose were really cool animals.

The other gift felt like a salad tray. However it was the Vogts gate street sign. I had known about this street while in Oslo. Vogt is a German name, a cognate of *advocate.* But the street namesake was definitely Norwegian. In the Oslo area it's a kind of blue-blooded family, and my wife appreciated some reactions she got when giving her full name. I can't remember what I learned about the *Vogt* of Vogts gate.

I assumed the sign I was given was legally purchased from the Oslo roads department. But it was actually stolen by two grad students who dressed up as street workers and, using a ladder, removed the sign from a building at the street corner.

They never got caught and many years later, one of them, Professor Annik Myhre, became department chair: I must have been popular if folks would steal for me!

It was in the early 2000s when Annik invited me back, all expenses paid, to be foreign 'opponent' at a PhD thesis defense. In Norway thesis defenses are public and formal. Yes, I wore a cap and gown. Rolf, the candidate, had his family in attendance.

I insisted that part of the defense was in Norwegian because I was pretty fluent.

Afterward, Annik said I was too easy in the questions I had asked. It's customary after the defense for the candidate's family to host a lavish dinner. The dinner was at a restaurant, you guessed it, on Vogts gate. I'm sure Rolf was thinking politically. I gave a funny toast in Norwegian: a high point in my career.

Annik and another staff member drove me back to my hotel. As we drove down Vogts gate, I asked where that sign had been. Annik pointed at the spot. Twenty-five years later, it still had not been replaced.

PART VII

Appendices

Like any proper scientific work, this semi-scientific narrative has appendices, four of them, adding detail to voyages; itemizing the terrestrial ports of my scientific life journey; and, naturally, offering a few References.

Kathy Crane: U.S. Ambassador to Norway Thomas Loftus; Peter Vogt; U.S. Naval Attaché to Norway; and Georgy Cherkashov discuss expedition plans on board the Russian research vessel *Professor Logachev,* 1996.
Photo: AIP Nielsen

Appendix 1

PhDs at Sea — An Interview

On board the RV Professor Logachev, *participants tell what's happening in a broadcast interview conducted by a Naval Research Laboratory historian.*

By Dr. David K. Van Keuren
History Office
Naval Research Laboratory

This is David Van Keuren, Historian with the Naval Research Laboratory. I'm speaking from the Russian Oceanographic Research Vessel the *Professor Logachev*, and I'm talking on 21 August, 1996.

This evening I want to discuss the results of the recent American, Russian, and Norwegian joint research cruise to the Norwegian Sea with those taking part: Peter Vogt, Kathy Crane, and Russian scientist Georgiy Cherkashev.

Dr. Van Keuren: Peter, why don't you start us off. We've now spent almost 28 days on the *Logachev*. We'll be going on 29 tomorrow. Any general reflections on how the cruise has gone?

Dr. Peter Vogt: I think we're all in agreement and certainly I feel very strongly that we've had an exceptionally successful cruise. I cannot recall having been on a research ship where such a variety of research was done. I would call this ship a floating laboratory, in fact. I am satisfied beyond all expectations.

We had a little bit of downtime with the equipment, but I would say, compared to my experience on US ships, it was at the same level or possibly less than what I recall, for the different types of equipment. I would say that a great advantage of having a ship that has so many research tools is that we could switch from one project to another quickly and make maximum use of the shiptime, so that at no time was the shiptime actually wasted.

We were able to arrive at a consensus here as to how to switch from one project to another. That was also a very good aspect that you don't have on other ships. My only comparable experience being on a ship with many scientists like this was the

Glomar Challenger. Twenty years ago I was on the drill ship, and this is the most nearly comparable experience.

Van Keuren: Kathy? Comments on the same topic?

Dr. Kathy Crane: We're still afloat. That's a good sign. We're going into Spitzbergen, I think with all of our projects accomplished. We were doing a lot of exploratory work, and had, by nature, to use a lot of different kinds of equipment because science is integrated into one whole.

I'm really happy with what we've come up with. I think we've got some really beautiful photographs of the sea floor, and great samples that will increase the number of rocks we've got from the mid-ocean ridge in this area [by] how much? One hundred percent? Probably.

Vogt: More than double the existing dredge collection.

Van Keuren: How does this compare to other oceanographic-cruises that the two of you have done? And I may also ask that of Gosha, who is sitting in on this.

Crane: I think I've been on over 30 expeditions so they've been really varied. Well, by far, there's a very large number of people on this ship. Generally, when I've worked in hydrothermal areas in the past, we've had a large number of scientists from many different disciplines. That's the nature of science. Biologists, chemists, geologists, physical oceanographers. And so, in that aspect, it hasn't been different for me. Maybe in the last few years I've been working on very small Norwegian ships where we've had a much tighter type of science, I guess. A narrow range of science being done on the seafloor. So I haven't really

410

done this type of work in 10 years, I guess, in a serious way on the ocean ridges. In terms of the international aspect of it, I think it's the first really tri-lateral expedition I've taken part in.

I've worked very often with many different ships from Italy, France, Russia, Germany, Sweden, Norway, and where else. So, that aspect, the collaboration, I think, has been fruitful on this expedition. I'm glad, from the very beginning, we had been able to set this up through people who worked together in the past rather than starting out cold without knowing the different partners.

Vogt: I'd like to make a comment about the Norwegian participation. This is trilateral, and, notwithstanding my best efforts, we were only able to get two individuals on board. One was a student and one a senior level technician, and they're wonderful people, and I'm glad they're out here. They did contribute a lot to the cruise, but to call it truly a trinational expedition would have been...

To really qualify for that term I would have liked to see a greater participation by Norwegian scientists. Maybe having four or five Norwegians instead of just two, but we tried and there were logistic and conflict problems involved so that we didn't have a bigger turnout. At the same time, as far as the US participation, I'm sure you've gathered by now that Kathy and I are, as we're getting a little older, spreading ourselves over greater areas of research.

We're becoming generalists, and a scientific generalist is very useful as a chief scientist because we have some feeling for these different disciplines, and we can synthesize, and we can say,

"Well, let's go over and use that tool in that particular area." But what we're lacking as far as the US team and the Norwegian team is to have some more specialists. That's my own feeling. Possibly Kathy and Gosha don't agree

I would like to, if we do a cruise like that in the future, to have some Americans come out with specialized pieces of equipment that would contribute in a laboratory setting and exploit the fact that the ship is large and has laboratory space. Then we could do analyses on samples together with the Russians. That's one aspect I would like to see happen.

Also, as you know Dave, Naval Research Lab has some very interesting geophysical and geochemical research tools. I could just name one: the DTAGS system. When it works, the Deeptow Multichannel System, which is the only one like it in the world, would allow you to get a high-resolution seismic image and the velocity structure for the top circa five hundred meters of sea floor. That would be a wonderful addition, and this ship would be the perfect vessel to host a system like that.

So, in the future, if we can find the funding, I would like to add to the set of research tools that we have out here and involve some more participants. As you know, all of you, I tried very hard to bring people in from our Marine Geoscience Division down at Stennis, where the focus is more on sediments than what Kathy and I do. I was unable to get even one person to come out, so that was a disappointment. It largely has to do with the funding situation that we have right now at the Naval Research Lab this year, which you're very much aware of. So I'm still extremely satisfied, and I can't see how we could have accomplished more than we did. But, nevertheless, I can see that the makeup of the teams could be a little more complete or representative.

Crane: Can I comment on that? I just think it would have been completely impossible to have done that this year because it was the seed money coming in to start a first major collaborative expedition with Russian scientists. From my perspective of raising money on the outside of NRL, they're willing to support the bare bones but nothing more, and if we were successful in pulling it off with a few number of American scientists [and] with expertise from Russia, then the statements were that [there was] much bigger likelihood, in the future, of bringing in a suite of experts. From my experience, that was the way it was, and we had no chance of raising the money for more scientists.

Vogt: We barely accomplished this. There were so many ifs, that we barely accomplished this.

Crane: By the skin of our teeth.

Vogt: An oddsmaker wouldn't have given very big odds for us to have succeeded a half year ago. I understand that our colleagues at Lomonosov and St. Petersburg were probably shaking their heads when we left. They didn't think it was very likely we would get this money over to them in three weeks and that we would actually go to sea with them.

Van Keuren: Comments Gosha?

Dr. Georgi Cherkashev: No. Maybe I'll comment about the scientific results later.

Van Keuren: We had a very rich scientific agenda on board starting out. There were several different research projects over several different types of areas. How much of that did we successfully accomplish?

Vogt: Well, having been the sort of primary initial instigator of the draft plan, I started out by proposing a number of research sites with the full realization that we wouldn't have time to do all of them. But, for practical reasons, it's good to have as many options as possible. Even if some sites have to be deleted, and, as you know, we had to delete some. We deleted three or four based on time constraints. We deleted one because the ice wouldn't let us get to the area.

The overall background for this research was the SeaMARC side-scan imaging, the reconnaissance imaging, that was done in 1989 and '90, as you know. Historically you are aware that—I think I told you before that in 1991 and '92—we, I and colleagues, tried unsuccessfully on several occasions to mount a major expedition to come here and ground-truth, meaning to go and look at specific problems like the object which we now know is a wonderful little mud volcano with all kinds of interesting anomalies.

We tried and failed a number of times to come out here. Most recently, two years ago, Chris Sandehoutier, a professor at Scripps who is into deep-tow technology, and I submitted a proposal to NSF to come here with the Scripps deeptow system and other tools, although not as many as the *Logachev* has. We never talked about having a complement of scientists out here doing analysis of methane and pore water and stuff. In this funding climate, the proposal got excellent reviews, but we got this typical letter back saying, "We get more good proposals than we can fund."

I, personally, have waited a long time, and this has been a wonderful fulfillment of six years of wondering what some of those things are on the seafloor. We've learned a lot, and one of the things that we've learned is that we have to be very careful about interpreting. We shouldn't be too quick about looking at and I'm talking about myself now seeing some little white lines down there and saying, "Ah! Those are sediment waves."

It turned out they were little crevasses in the seafloor, and we didn't have the capability with SeaMARC to look at the topography associated with those patches. Had we investigated them in detail with other ships we would have known that.

Another area where we went in, the Lofoten Basin, we also discovered things that we didn't expect. It was counter to our expectations. We found that there were places there which we thought were soft-sediment waves, and we found it was impossible to even take sediment cores. I think we all agree that our premier accomplishment was the investigation of the mud volcano, and I think the fact that the sonar imagery was so exciting and showed all these new features, in that we found a temperature anomaly, which cost me a case of champagne, in the water above the mud volcano, which is probably a historic first: Nobody has ever, over a so-called cool or warm seep, found a temperature anomaly. If it stands up to analysis— I guess I concede that it does—I will buy the case of champagne.

Then finding the methane in the water column and the extremely high heat flow in the center: The gradient is five degrees per meter, which is a stunning temperature gradient for a passive continental margin. So there were many exciting aspects that were not entirely predictable. We had little clues from last year, but now we know a lot more. I think, probably, if you took a vote on the ship you would say that, by far and away, the Haakon Mosby mud volcano was the most exciting thing we looked at, but some of the other problems were also very exciting. Okay, I've talked enough.

Van Keuren: Do you agree with this Kathy?

Crane: About which areas were the most exciting?

Van Keuren: The portion of our agenda that we actually accomplished. What do you think are our major accomplishments scientifically?

Crane: Actually, I was surprised when I came on the ship. I sort of missed out on the last couple of weeks of planning. I was doing other work, so we ended up doing about a 120 percent of what I originally thought we were going to do. So I wasn't prepared for some of these other stations in the south, which added on to the variety of things. Yeah, I think we've had marvelous success. I'm really glad we were able to bring color photography to these issues.

I think that we discovered a lot of very really interesting features about this region along the Senja fracture zone between the Barents Sea and the Norwegian/Greenland Sea. Along the Knipovich Ridge there are some really interesting areas. We have some enticing pieces of information that may lead us to come back and do further studies in particular areas. So, I think we had a lot of success.

Van Keuren: In your eyes, what one accomplishment on this trip would make your colleagues back home pay attention.

Crane: I would say the methane hydrates frozen on the seafloor. You bring it up and it's gone. You don't know it's there. So, we took all these photographs, you see. Beautiful contrast, the whites against the dark. The frozen hydrates are just incredible. Pictures are a thousand words often, so one photograph of that plus a beautiful side-scan sonar image of the mud volcano. It is just stunning. I think a couple of those things together would really be a wonderful eye opener to a lot of people.

Van Keuren: What about the Knipvoch Ridge?

Crane: We had some debates about it. I actually think we found bacterial mass. There's a question whether or not it's a developer, but there were a couple of frames on both rolls— the Russian roll and the US roll of color film at the same time. They were developed separately, and they had the same pattern on them, so Pete's going to check this out. They're in an area of subtle temperature anomalies, and the topography is so rough there it would really take us at least two weeks to a month to really investigate that area thoroughly. We just had such a quick run through. Normally with mid-ocean ridges when we are investigating them, we would spend a whole month in an area like that. Because it's such complicated photography, and plus it's fascinating for me to see the basalts. Some areas which we thought were very fresh but had a dusting, more than a dusting, of sediment on them. But they were like fresh basalts in the sediment. So I think it's some kind of different terrain, and there's a lot. Knipovich Ridge doesn't reveal its secrets easily,

I think. So it will take us some time to really go there and investigate it with a whole bunch of instruments.

Vogt: Yeah, we could easily spend five successive cruises looking at the ridge system. I like to think back to exactly 30 years ago now when I was a student on board the navy icebreaker *Atka*, previously the *Southwind*. I think it's one that the Soviets had during the Second World War. Then it was given back to the United States because they still had some plates and cyrillic lettering on the *Atka*. We worked in the Barents Sea, and on the way back we took a couple of dog-legs, the first tier of a polygon. I crossed the Knipovich Ridge rift valley, and I stuck them into my thesis, and at that time we thought we had discovered something new, and we called it the Atka Ridge. Later we found out that the Soviets had already discovered the feature, and so the name Atka Ridge was not to last.

But I think back, looking at those profiles, and this was before plate tectonics had been formulated but after Vine and Matthews. At the time, we argued what the function of this linear feature was. Whether it could be a fracture zone, or a spreading ridge, or even a trench. You know, our hypotheses were all over the map. For me, it's kind of historic to come back, virtually to the same place, 30 years later.

Crane. We should have fireworks.

Cherkashev: Maybe some additions about the mud volcano, that unique place. As I know, this is the first time we've found huge surfaces where gas hydrates are on the surface of the sea bottom. We made a very good record there and did a lot of interesting things. As for Knipovich Ridge, this is the first step, but a very successful step because it was a surprise for me that we found the temperature anomaly. Finally we found the mineralization—low temperature but nevertheless hydrothermal mineralization. So it is a signature that the rock must be here making a high temperature. This is a very good result.

Van Keuren: Any major disappointments?

Vogt: No, the only disappointment I have is that even though we've been out at sea now for 30 days I still haven't really got to know a lot of the scientists well, and that may be partly, speaking personally, a language problem and partly because there are many people, and we try to work with the people that we normally work with. They're a part of our working group. So, if I had any disappointment it would be that I didn't get to know the other scientists better than greeting them in the passage ways and so forth, but maybe that's for the future to correct.

Cherkashev: It's a little bit of a pity that in the last stage, the last part of our voyage, there was a problem with the sonar, and we couldn't have this data for a big section of the Knipovich Ridge. We studied only a short segment, but this is for the future.

Vogt: When we write a proposal and put in the sample of the deep-tow side-scan and some of the pictures, I think we will be miles ahead and the people will want to pay us, I hope, to image the rest of it with that kind of resolution.

Crane: Disappointments?

Cherkashev: Kompot? (He offers a fruit juice that is a staple in Russia and Ukraine.)

Crane: Not for me. For the interviewer that was a major disappointment. I actually like Kompot.

Let's see. Major disappointments? Nothing beyond the normal disappointments of going to sea and having things break down, but that's normal for me. There are so many people on this ship and 30 days, almost 30 days, went by really fast for me. There were a lot of people I just didn't have a chance to go to. Seminars were always during the time I wanted to fall asleep the most. During watch and after watches. I think we did the best we could in the situations. Good opportunities for people to mingle and mix every now and then. I just see it as a stepping stone toward further collaboration. I, personally, would have liked to have seen a lot more photographs of the seafloor.

Vogt: Can't get enough.

Crane: Yea, can't get enough photographs. I'm always surprised that what you really see on the seafloor, and how it's often different from what you imagined for years and years about being there. I have always had this experience in diving in submarines. When every time we actually go to the sea for a week, we discover new things visually. It has a real impact on me.

Cherkashev: It was a moment in the evening that we had a video tape of basalts and very strange forms. It was a discussion if it was basalts or maybe it was sulfites. So it was a very dramatic situation because we hadn't yet the color photo, so it was not possible to recognize for sure if it is sulfites or not. We decided to have a dredge there. Finally we got basalts, and it was a little bit of a disappointment. When it was finally dated it was not a sulfite.

Crane: You were probably right next to the vent.

Vogt: I don't think we proposed, or anybody expected, to find a black smoker with the tools we have on board. It was the first step in prospecting. We probably got more than we expected. I didn't think we'd find even any temperature anomalies. So we were already ahead. We found the smoke from either a big gun or a cap pistol.

Crane: A couple. One on the volcano and one on the Ridge.

Cherkashev: It seems that we were very near, very close to this hot vent. It was maybe this seamount near the Valley. Now, I can say that maybe we should work more there, but we concentrated in the valley where we could take the channel, but the origin was in the slope on this seamount. Maybe next time we have an assignment we'll discover it there.

Vogt: Well, for one week...It was really a lot accomplished just for a single week.

Crane: There is an example of when we did the first long tows along the mid-ocean ridge in '83 and '84. I think it was the first time anyone tried to prospect for hydrothermal features, along the ride back when we crossed it. We first started it on the Juan de Fuca Ridge and later on the East Pacific Rise. It took us about an entire month to really do it well with side-scan and temperature anomalies. Then we patched everything together, so it's a lot longer period of time to work on.

Vogt: We really should propose to spend a month and just tow the side-scan from south to north and then north to south again with the CDT attached and just not bring up the sonar as long as it works. We'd just keep on towing it. So, Dave, if you know of any funding sources you're welcome to take any materials that we have and go around to funders and whet their appetite and tell them where we can be found. Pass our email and fax numbers out to them.

Van Keuren: How did the American and Russian scientific teams mesh together?

Cherkashev: Without any problems.

Crane: Especially at ping-pong. I think things went fairly well. There were some areas where we didn't have comparable people working in the comparable fields. But when we were working on the deck and box-core sediments there was some expertise that we had like heat flow or the CDT work or other expertise the Russians had but which didn't overlap too much. Personalities did just fine.

Cherkashev: I remember that before, during the period before, we spoke of whether it is good or not to have specialists in the same fields on a ship. We agreed that it would be good, it would be more good for cooperation. So I think that in the future we should do it.

Van Keuren: Is there a difference in what I might call the scientific styles in which Russians and Americans work at sea? There has been a lot of discussion in historical sources of scientific styles. Particularly how scientific styles differ between countries. Did we see any of that here? Do Russians and

Americans pursue oceanography, particularly oceanography at sea, in different ways? Did any of that manifest itself during our cruise?

Crane: I don't know. Maybe all the expeditions I've been on, that are not on small ships, we had a very strict watch schedule where we would take whoever was there and break people up so they all worked together on one watch rather than in separate teams. So that we'd have things covered round the clock, we'd all work together. It's my impression that the Russians operate things more in teams rather than having pieces of teams broken apart so you have members of different teams working together all on one watch for four hours. I think there's probably some little differences there, I don't know, maybe because people spend a lot more time at sea in Russia.

Cherkashev: It's only in this situation because we have a lot of teams here, from Moscow, from Lomonosov Polar Expedition, from our institute, and from Moscow from three institutes: from Academy of Sciences, from Shirshov Institute, Microbiological, and Geochemical Institute. So it's a mosaic. It was not very easy to do it, but finally it has been a success.

Vogt: Normally, we want to do more.

Crane: Working more on the deck.

Vogt: Working more and not necessarily being the primary deck or winch operator. I've never operated a winch except for an old ex-PT winch, but still I felt that everything was done, which was great. I was kind of lazy on this ship, but everything was done.

Crane: But you found that frustrating, too.

Vogt: A little bit. You feel ambivalent about it. I felt like a bystander, like I had suggested the problem and then everybody covered all the fields. You know, when we go out we swizzle the knobs much to the dismay of the technicians. We go to the graphic recorder, and we change the rates, and we fiddle around with the tuning, and often we end up producing a worse record than if there were a single technical person responsible.

So I have to say from a purely technical point of view, taking, for example, the profiler, which I was interested in—and it's the type of equipment we run continuously—to have Peter Kiritsky's group responsible for it, and tuning it, and kind of keeping us away from it had the benefit that the quality of the data is probably better than it would have been if I had fooled around with it. I've heard that complaint in the US often from the technical people, that the scientists come in, and they take over, and they swizzle the knobs, and they try to get this better and that better, and there are so many variables, and we don't have the experience. Kathy may disagree. I know I have fooled around with it, but often somebody who has devoted their life to a piece of equipment is in a better position, even though they are not geologists, to tune the instrument to get the best possible record. Particularly if they have a little direction from the scientist. So there's two sides to that, like I've gone in and fooled around with the record.

Crane: I do have mixed feelings about it. As students in America, I don't know whether it's right now, but it was our job on the watch to change all the records, do everything from the ground up. So you really have to understand how profiles are made, what they mean. Your job was to sit there to make sure that the profiles didn't run off the page, and log everything. That was our job.

Vogt: Sure, but you wouldn't have done as good a job as I've seen what watchstanders do on our ships and with very few exceptions.

Crane: It depends on what you want. If you're in a training mode of coming out to train oceanographers to have a real understanding of what it is they're looking at, you'd be someone who's always been exposed to everyone else fixing the records and changing them and would have no idea even what a change in scale is. But that's my opinion.

Vogt: We had the benefit of having gone through that phase where we had to change the paper and fiddle around, but for a student to come out who didn't have to do that it is possible that they would have less understanding of it.

Crane: The same thing goes for transponder navigation. I had to do all the navigation, and we had to launch all our transponders, and we had to navigate them in as scientists. We had to do all that work. So you had to complete, we had to learn all the technical aspects before we could even interpret the geology. Then, on the other hand, it really helps you if you're planning a survey and someone says, "Should you use transponders here? Where should we put them?" Maybe you have in your mind different scientific objectives than an engineer would have.

There are positive attributes to both areas. I personally enjoy going to sea to work with equipment and to fly instruments. That's why I became an oceanographer; otherwise I would have stayed at home. Most American ships would keep records going all the time through the whole expedition. Would never turn anything off. You would collect data from the very time you left port.

Vogt: Because you never know where it's going to be important.

Crane: Ship time is so expensive. It's $12,000 a day in America right now.

Vogt: But I can see the fear of equipment breaking down and not having spares. That's probably the driving reason why we had to twist arms to get the recorders running all the time. My experience has been different from Kathy's in some respects. There are things we've done on the ship that I have never been a part of. For example, the transponder work. I have been on ships that have done side-scan, of course, but I have not ever been on a ship before that did camera sled runs. I've been on a ship where we did photo type, you know, single pictures, but even that I have very little experience with. So my experience has been more with cruises where we did mowing-the-lawn type surveys, and also I've done some things in the field like going out on an aircraft. We come at it from a little bit different perspectives.

Crane: I did aircraft work in Kenya.

Vogt: She did the airborne heat-flow work. They'd drop heat flow probes and circle around until...

Crane: No I didn't (laughs). I flew over Masai in Kenya.

Vogt: Kathy's done everything. It's hard to beat her.

Crane: No, I haven't done everything.

Van Keuren: Is there any way you could pinpoint in which, as a result of this being an American and Russian cruise, it has differed from what a purely American cruise would have been like with the same objectives?

Vogt: We've already touched on some of that, I think, partly for economic reasons, we are unable to bring...I say, for economical reasons we would not have a ship that is a floating laboratory like the *Joides Resolution* or that the *Glomar Challenger* was, and this ship is. Because our ships aren't big enough. That's part of the reason but also the high labor costs, the high insurance costs, the high ship costs. Everything is so prohibitive that we tend more to go out and do a specific thing like running the ship back and forth, pulling along the Seamark, without having a large number of people aboard. So that's the difference.

Crane: On American expeditions, people came for a month, then left at the end of the month. People flew into port, and then flew out again after a month. Then a new team came in. Imagine people riding the ship for six months.

Vogt: That's a big difference there.

Crane: You need a big ship.

Vogt: It becomes more like a home, like the Russian- or Soviet-era ships were more like home. They stayed out for months, even up to a year at a time. It would be unthinkable for an American ship. There are union laws that prevent that. I think, now the typical cruise is a month, or 28 days for US ships. So, that's a difference. From a cultural point of view, I haven't experienced as much cultural life on US ships. I think people tend to be more insular, and they stay in their staterooms. Your experience has been different.

Crane: My experience has been not to have any parties or have very little in the way of parties or the cultural things we've had, like giving Russian lessons and things like that. It makes the life out here more complete. And the lectures. If you have enough scientists, then you can have lectures and that's a new experience. Only on the *Glomar Challenger* did we have scientists give talks because there were enough. My experience has been, going to sea, I'm the chief scientist, and I have four or five or six people working with me, and there were no, maybe one or two, people that could be called a scientist, and so we weren't giving lectures, and weren't having parties, or any of it.

Crane: On a Navy ship?

Vogt: Well, yea. NAVOCEANO and NRL ships and so forth. Also I mentioned this before: NRL, for economic reasons, lost its last ship in the year 1982. So in a way, we were in a similar kind of boat, if I can use a bad, ironic pun there, as our colleagues in Russia. Starting already—not now, not 1996, but 14 years ago—we no longer had our own ships, so we had to start working on other people's ships, which also makes a big difference if you have your own ship or not.

Cherkashev: You said about the seminars. I remember there is an educational aspect of our expedition. We have a team of students, an international students team of 10 persons.

Vogt: Yea, total.

Cherkashev: It's very important for them, for their future work and for science in Russia and Norway.

Vogt: Yes. Absolutely so. We not only have a floating mini-laboratory, we have a floating mini-university. The *Logachev* University.

Van Keuran: I'd like to throw that same question back at Gosha. Has the presence of Americans on this ship in any way changed what it would have been like if it had been a purely Soviet, or let me say Russian, trip.

Cherkashev: Not a lot. It was a usual atmosphere on the cruise, but usually there was some influence from the guests. It is traditionally in Russian's character: If you have a guest the best for them and for us. It was the Soviet time, and now too, foreigners for us was some exception. But now when the differences between us are less and less this is not an unusual situation for us. We work together with Finish specialists and with Americans and French, so this is not very different than if it were a pure Russian cruise. It's more concentrate of science because it's very good that we enrich each other with our knowledge. This is very important to us, this cooperation.

Van Keuren: Will we see more of this sort of international scientific cooperation? Particularly between Russia and America?

Cherkashev: Sorry?

Van Keuren: Will we see more of this type of international cooperation between the Americans and Russians?

Cherkashev: It depends, but I hope it will be. It must be.

Van Keuren: Kathy, Peter. Any comments on that?

Crane: Well, I hope so. I've been working for four years trying to build Russian-American collaborative programs. I think it's extraordinarily important for world peace. That's why I'm really interested in this, this whole effort. I don't approach this just as science objectives. To be quite honest, I'm more interested in the international relations aspect of it, between Russia and the United States. It's far reaching. So, I hope that in this next year, I intend to work on getting follow-up expeditions that will be collaborative programs going between us.

Vogt: It would be great if it were possible to set up a longer-term collaboration so we don't go through all these machinations for just a single cruise, but we[would have, for example, a five-year plan, and we've planned a number of cruises. Maybe one every other year, maybe one every year, maybe a smaller one and a big one, and there's some continuity to it instead of a one-shot effort like this. That would be my wish. It would be terrific if we could convince funders to do that.

Van Keuren: What follows on after the cruise of the *Professor Logachev?* What's next?

Cherkashev: For the ship?

Van Keuren: For the ship.

Cherkashev: For our team?

Van Keuren: For the team.

Cherkashev: For the team. I hope they'll continue our relations, and we'll begin to process this data which we received, and we have some plans for the near future and for next year, not only connected with this material but new research.

Van Keuren: Can you get me more details on this?

Cherkashev: Maybe Peter can say better.

Vogt: Well, in the first place what we plan to do is work collaboratively on the data and, as you know, the final product should be a series of co-authored scientific publications in the best scientific journals that are most widely read. That's objective No. 1. What we would also like, what we will probably start to work on right away, is planning for some kind of a workshop which we might host in the United States. It would be logical to have it in Washington because most of us are located there. So we have tossed it out, first as a joke, and then we thought more seriously about having a workshop maybe next year in connection with the American Geophysical Union meeting in Baltimore and to have some Russians from this expedition go there and present their results.

I, half in jest, suggested that the ship should come across the Atlantic to the United States. I think that would be a great thing if we could pull that off. I guess I'm drunk with success. We were able to do this, so I figure if I can do this, we can get the ship over there, which would solve some other problems, like the cost of putting people up and the transportation. If we could think of some creative, fundable project for them to do in the Atlantic, we might bring them over as a part of the research cruise in the Atlantic. Again, I don't know if we can do it in time for the next AGU. It's less than a year from now. Something like that would be a terrific coup. Of course, we would plan to take some trips to Russia also, Kathy and I, to follow up.

Van Keuren: (to Kathy Crane) What's next?

Crane: Funding. Right now we have a program, Gosha and I have a proposal, for collaborative work-up of the data. It's through a program called—I forget what it is. Civilian Research or something.

Vogt: CRDP.

Crane: Defense Foundation. Money particularly to go to Russian scientists working with us. Whether or not that's funded I don't know. We should find out soon. Another plan we have would be the use of Russian submersibles, the Mirs, on certain of these targets in the next year, and I know they are very interested in this. Whether or not that would involve some very interesting juggling of money, raising money from, I think, several different countries and organizations. That's what I can perceive right now as being follow-up. Then we also have this additional program in which many people here are involved.

That's the International Arctic Environmental Program, which is not related to this, but there's overlap between the institutions. That may, or may not, have on-going funding within the next year, and, if so, it would have a major impression on my salary, and I know a lot of other Russian salaries could benefit from that, too. A lot of these things are just up in the air. Election year in the US and directions—you don't really know which way things are going. Maybe things will have resolved themselves this last month while we have been gone, I don't know. But we have made efforts before the cruise for continuing funding. We'll just have to wait and see, I think, right now.

Vogt: Then, we're also considering the possibility of coming out here to collect deep sea ooze and marketing it in New York City as a medicinal ointment. Dr. Kathy Crane has plans to have this material tested for its toxicity. There are rumors around that we can deny and therefore make them vettable. It will prolong life and virility, and maybe we can get the Chinese to buy deep sea clay from us instead of shooting rhinoceroses for their horns. So, those are my closing thoughts.

Crane: Fifty dollars for five ounces, I'm sure. No problem.

Van Keuren: Any other closing thoughts?

Crane: It's been a lot of fun for me. Normally when we go to sea on American ships, it's just so exhausting because you have to do everything. We never have parties and never have any fun at all. So I just feel exhausted after an American expedition and during them. Before, during, and after.

Vogt: Not to speak of the *Haakon Mosby*, I do have one final thought: Not all oceanographers have iron stomachs. There are many times in my life when I have regretted ever becoming involved in ocean research, but when the storms end and you go back you forget all that stuff, and then you go out again.

Crane: Except on a Norwegian ship. Particularly the *Haakon Mosby*, which is the ship we used for the SeaMARC work. We had to pay the dues to collect those data, the SeaMARC data, which really formed the basis of this followup.

Cherkashev: We were very lucky with the weather. It was a gift from Neptune, I think, for us, for our cooperation, for the science. We should continue our efforts.

Van Keuren: Thank you very much.

APPENDIX 2

Trip Report:
RV Keldysh from
the Norwegian Sea
summer 1998

Trip Report

Gas hydrate cruise on RV Keldysh in the Norwegian Sea, June - August, 1998

Dennis Lindwall
NRL code 7432

This was an international cruise involving 50 scientists and technicians from at least 5 nations and 13 institutions. The cruise went to 4 sites in the Norwegian sea area with all but one site being a gas hydrate site. These 4 sites were the Storrega Slide, the Haaken Mosby mud volcano, the Knipovich ridge, and the Vesnessa ridge. Multiple submersible dives were made at all of these sites and some additional shipboard work was done at these sites. This cruise will get a lot of public exposure from the two television crews and two magazine photographers (from the National Geographic) who photographed both the scientific aspects and some of the personal aspects of this cruise. NRL should exploit the potential positive benefits that we could get from this exposure.

MIR-1 being deployed from the Keldysh at the Knipovich Ridge on June 20, 1998.

Scientists from 5 nations discuss results from the Storrega slide and plan the Haaken Mosby mud volcano dives.

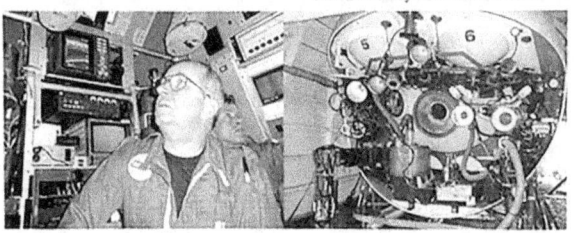

Anatoly Sagalevitch, the head of the deep manned submersibles laboratory of the Shirshov Institute of Oceanology in Moscow. He was the pilot of MIR-1 for the first dive.

The front of MIR-2 showing the view ports, lights, manipulator arms, slurp gun hoses, sample collection trays and other scientific instruments.

APPENDIX 3

Waypoints in a Half-Century of Exploration

with dates and purpose of visit

*Guide to Abbreviations
    A: by air
    C: conference, workshop other business
    ECL: Eclipse cruise lecturer
    M: peaks named or climbed
    NV: never visited
    S: visit by sea: port stops, embark or disembark
    T: shore-leave tourism
    U: education
    V: viewed from ship

AFRICA
Abidjan, Ivory Coast (A: 1968)
Capetown, South Africa (A: 1969; S: 1987)
Dakar, Senegal (S: 1975)
Fernando Poo (now Biroko), Spanish Guinea (A: 1968)
Johannesburg, South Africa (A: 1968)

ANTARCTIC
McMurdo (Oct. 1962–Feb., 1963)
Vogt Peak Geologists' Range MNV: (shares my name)

MIDDLE EAST
Aden, Yemen (S: 1964)
Beirut, Lebanon (S: 1964)
Suez Canal (A: 1964)
Zerbirget (aka Zabargad, St. Johns Island), Egypt (S: 1964)

ASIA
Singapore (A: 1968)

ATLANTIC ISLANDS
Ascension (V: 1975; A: 1984)
Azores, Ponta Delgada (S: 1975)
Bermuda (A: 1969; S: 1975)
Canary Islands: Las Palmas, Gran Canaria (S: 1975)
St. Helena (S: 1975)
Tristan da Cunha (S: 1975)

AUSTRALIA
Adelaide (A: 1968; C)
Canberra (A: 1968)
Darwin (A: 1968; S: 1972)
Perth (A: 1968)

BRITAIN
Ben Nevis, Scotland (M: 1965)
Eddystone Light (V: 1964)
Edinburgh, Scotland (T: 1966)
Fort William, Scotland (T: 1965; C: 2004)
Newcastle (S: 1965)
Plymouth (S: 1964)

CANADA
Calgary (C: 1981)
Halifax (C)
Ottawa (C)

CARIBBEAN
Martinique (ECL: 1973)
San Juan, Puerto Rico (ECL: 1973)
St. Lucia (ECL: 1973)
Trinidad (ECL: 1973)

EUROPE
Berlin, Germany (C: 1980)
Bressanone, Italy (C: 1980)
Ceuta, Spain (S: 1964)
Hamburg, Germany (C: 1968)
Innsbruck, Austria (UT: 1961–'62)
Kiel, Germany (S: 1993)
LaSpezia, Italy (ST: 1964)
Madeira, Portugal (V: 1964)
Madrid, Spain (C: 1970)
Monaco (S: 1964)

Paris, France (C: 1984)
Rostock, Germany (C: 1993)
Straits of Messina (V: 1964)
Stromboli, Italy (AV: 1964)
Viareggio, Italy (V: 1964)
Vienna, Austria (C: 2007)

ICELAND
Reykjavik (C: 1974)

INDIAN OCEAN
Diego Garcia (A: 1990)
Mauritius (S: 1964; A: 1868)
Prince Edward and Marion Islands
          (NV: 1979 Vela-Israel nuke test)
Reunion, Seychelles, Victoria, Mahe Island (S: 1964)

INDONESIA
Djakarta (S: 1971)

LATIN AMERICA
Cotopaxi (volcano), Ecuador (M: 1960, '63)
Montevideo, Uruguay (S: 1987)
Panama (S: 1974)
Paramarido, Suriname (A: 1968)
Orizaba (volcano), Mexico (M: 1960)
Popocatepetl (volcano), Mexico (M: 1960)
Recife, Brazil (A: 1968; S: '85)

NEW ZEALAND
Auckland (C: 1971)
Christ Church (S: 1962, '63, en route to and from McMurdo)
Wellington (C: 1972)

PACIFIC ISLANDS
Galapagos (SV: 1974)
Guam (S: 1972)
Hawaii: Oahu (A: 1968)
Vogt Guyot (NV: far western Pacific seamount shares my name)

PAPUA NEW GUINEA
Port Moresby (A: 1968; S: '72)

RUSSIA
Arkhangelsk (C: 1992)
Cape Chelyuskin, Siberia (V: 1965)
Moscow (C: 1992)
Novaya Zemlya (V: 1965, '66)
St. Petersburg (C: 1995)
Vilkitsky Strait, Siberia (V: 1965)

SCANDINAVIA
Bear Island, Norway (V: 1965)
Bergen, Norway (S: 1966, '89, '95, '96, '99, 2010)
Copenhagen, Denmark (S: 1965, '85; C: 1978)
Longyearbyen, Svalbard (aka Spitsbergen)
        (SV: 1988; V: 1995, '96)
Oslo, Norway (S: 1965, '77; U: '78; C: 2009)
Tromso, Norway (S: 1995)

UNITED STATES

Annapolis (S: 1994, '95, '97, '98, 2005)

Andrews Airforce Base (now Joint), Maryland (A: 1969)

Baltimore (C: 1996, 2010)

Barrow (Utqiagvik), Alaska (C: 2025)

Bay St. Louis, Mississippi (C: 2000)

Boston (A: 1969)

Caltech, Pasadena, California (U: 1957-'61)

Cape Cod, Woods Hole Oceanographic Institution (S: 1964)

Denver (C: 1988)

La Jolla, California, Scripps Institution of Oceanography
        (S: 1958)

Madison, Wisconsin (U: 1963–'67)

Norfolk, Virginia (S: 1975, 2000)

Naval Air Station Patuxent (A: 1984)

Pittsburg, Pennsylvania (C: 2003)

San Antonio, Texas (C: 1986)

San Francisco (A: 1968; C)

Washington, DC (S: 1970; C: '65+)

APPENDIX 4

References

Bates, C.C. 2006. *Hydro to NAVOCEANO: 175 years of ocean survey and prediction by the navy.* Rocktin, IL: Corn Field Press.

Berlitz, C. 1973. *The Bermuda triangle.* New York: Avon Books.

Bowin, C.O., and Vogt, P.R. 1966. Magnetic lineations between Carlsberg Ridge and Seychelles Bank, Indian Ocean, *Journal of Geophysical Research*, 71(10), 2625-2630. (21 citations)

Crane, K. 2003. *Sea legs: Tales of a woman oceanographer.* New York: Basic Books.

Glen, W. 1982. *The road to Jaramillo: Critical years of the revolution in earth science.* Stanford, CA: Stanford University Press.

Heezen, B. C., and Hollister, C.D. 1971, *The face of the deep.* New York: Oxford University Press.

Hurly, R. 1968. Southwards with Project Magnet, Astronomical Society of Southern Africa, 6., p. 53054.

Johnson, G.L. and Vogt, P.R. 1973. Mid-Atlantic Ridge from 47° to 50° N, *Geological Society of America Bulletin*, 84(10), 3443-3462.

Kious, W.J., and Tilling, R.I. 1994. *The dynamic earth: The story of plate tectonics.* US Geological Survey.

Maury, M. F. 1856. *The physical geography of the sea.* New York: Harper and Brothers..

Petrov, R. 1968. *Across the top of Russia.* New York: David McKay Co.

Petrov, R. 1965: 15 July, 18 July, 18 August, 17 October, and 23 October. *New York Times.*

Simkin, T., Tilling, R.I., Vogt. P.R., Kirby, S.H., Kimberly, K., and Stewart, D.B. 2006. This dynamic planet: World map of volcanoes, earthquakes, impact craters, and plate tectonics, geologic investigations map I-2800. US Geological Survey.

Strangway, D.W., and Vogt, P.R. 1970. Aeromagnetic tests for continental drift in Africa and South America. *Earth and Planetary Science Letters,* 7(5), 429-435.

Tucholke, B.E., and Vogt, P.R., eds. Initial Reports of the Deep Sea Drilling Project, v. 43. Washington, DC: US Government Printing Office.

US Naval Oceanographic Office Geomagnetic Surveys. 1970. Informal Report, UNCLASSIFIED, approved for release by G.A. Yang for H.P. Stockard, Director, Magnetics Division.

Vogt, P.R., and Ostenso, N.A. 1966. Magnetic survey of the Mid-Atlantic ridge between 42° N and 46° N, *Journal of Geophysical Research,* 71(18), 4389-4441. [44 citations]

Vogt, P.R., and Ostenso, N.A. 1967. Comments on mantle convection and mid-ocean ridges, *Journal of Geophysical Research,* 72(8), 2077–2084. [Only 4 citations as of Feb 2024!]

Vogt, P.R. and Tucholke, B.E. 1986. Imaging the ocean floor: History and state of the art, Ch. 2 in Vogt, P.R. and Tucholke,

B.E., eds., *The Western North Atlantic Region, v. M in the Geology of North America*, The Geological Society of America, p.19–44.

Vogt. P.R. 1986. Magnetic anomalies and crustal magnetization, Ch. 15 in Vogt. P.R. and Tucholke, B.E., eds. *The Western North Atlantic Ocean Region, v. M in the Geology of North America*, The Geological Society of America, p. 229–256.

Vogt. P.R., and Jung, W-Y. 2000. GOMaP: A matchless resolution for the new millennium, *EOS*, 81 (23), 254-258.

Wegener. A. 1928. *The origin of the continents and oceans*, 1961. English translation from German by J. Birham. Mineola, NY: Dover Publications.

Wertenbaker, W. 1974. *The floor of the sea: Maurice Ewing and the search to understand the Earth*. Boston: Little, Brown and Co.

Acknowledgments

A multitude of folks shaped my career. Most are long gone, but my gratitude endures. I was introduced to glacier ice by Herfried Hoinkes (Professor, University of Innsbruck; 1961–'62) and thus hired by Bob Black (Professor, University of Wisconsin) to help his Antarctic studies (1962–'63). With some ice and 1964 ocean experience, I was accepted by Professor Ned Ostenso not only as as a master's and PhD student but also sent to the European Arctic on icebreakers (1965–'66). Ned also convinced me to accept the offer by NAVOCEANO recruiter Dr. Charlie Bates to move to the DC area when I had instead hoped to head to academia out West. In 1975 when NAVOCEANO was 'pork-barrelled' to Mississippi, I flirted with academia but thanks to Henry "Hank" Fleming and Dr. John Goodman of Naval Research Laboratory found a research position there, so beginning my long commutes to DC. Thanks to NRL Director of Research Dr Alan Berman for approving my year as visiting scientist at the University of Oslo, and to Norwegian professor and friend Olav Eldholm for the invitation and locating a house for our family of four to rent. Among many other Norwegian colleagues our friend Eirik Sundvor (Professor, University of Bergen) also stands out. I have collaborated with many other scientists, among them many coauthors. Most notable is Dr Brian Tucholke of Woods Hole. We were co-chiefs on Leg 43 (1975) on Glomar Challenger and co-editors of 1986 DNAG volume M on the Western North Atlantic Region. I fondly recall collaborating with Tom Simkin (Smithsonian Institution), Bob Tilling (US Geological Survey) and others on the 1994 and 2006 This

Dynamic Planet global maps. Research of what underlies the Chesapeake got me to know and collaborate with USGS Tom Cronin and Deb Willard, and Jeff Halka of the Maryland Geological Survey.

Dabbling in Calvert Cliffs and Southern Maryland geoscience over many years brought collaboration, still ongoing, with paleontologists Ralph Eshelman and Stephen Godfrey. Today in my octogenarian dotage I collaborate via email with Gillian Foulger (Professor, Durham University, UK).

Karma, chance, or Lady Luck: Such a force has to be acknowledged as my career took off in auspicious directions. I had not planned to study the ocean floor even as a beginning graduate student. But one day while chatting with another student, Frank Birch, I learned about a planned 1964 Indian Ocean expedition from Woods Hole and the opening for a research watch stander. My career was thus influenced by the many scientists, students, crew, and interesting folks I met in ports. That cruise further persuaded my Wisconsin professor, Ostenso, to send me to sea on icebreakers.

My auspicious move from near DC to Southern Maryland was also due to chance. Eric Schneider, who recruited me into his new GOFAR division at NAVOCEANO, was looking for empty office space to house the group. His new illustrator, Barbara Grosvenor, was living on the Randle Cliffs NRL base south of Chesapeake Beach. There was office space there.

I especially thank my wife Randi and two sons for putting up with long absences and our joint efforts to help keep Calvert County the special place it is.

Thanks to my long-time research colleague Dr Kathleen Crane for her gracious Foreword. And to Fred Bowles, another GOFAR and later NRL fellow geoscientist; Jim Borell; Charles "Chip" Martin, and famous paleoclimatologist Bill Ruddiman for their early reviews.

Finally I thank New Bay Books publishers and now friends Sandra Olivetti Martin and William Lambrecht, and designer and artist Suzanne Shelden for their endurance assembling all these fragments into a coherent book.

Photo Credits

Cover:
High rolling seas toss the RV *Chain*. Photo by Peter Vogt, who promptly closed the hatch and retreated after his shot.

Back Cover:
The icebreaker *Atka* takes a break from work in the Arctic Ocean to offer ice leave and beer amidst the serenity of the frozen ocean. Photo by Peter Vogt.

P. vi:
Peter in Scotland in 1965 along famously picturesque Loch Linnhe, which stretches 30 miles, from Fort William to the Firth of Lorne.

P. 1:
Peter and Tom Berg (left) on Ross Island install a Wincharger generator to power weather-monitoring instruments. Berg died in Antartica in 1969 in a helicopter crash. Photo by Professor Robert Black, who was in charge of the project.

P. 84:
On an ice floe, Peter deploys an instrument for measuring the force of gravity.

P. 147:
On Bartlett, Peter confers with Australian geologist John Connolly, who was teaching at the University of South Carolina at the time.

P. 302:
Aboard *Keldysh* with Mir pilot (center) and engineer (right) in dive suits ready to submerge.

P. 374:
The *Marion Dufresne*, one of the world's largest research vessels, detoured into Chesapeake Bay to take core samples documenting the changing depths of the Chesapeake since the last ice age.

P. 404:
Peter reviews expedition plans aboard the Russian Oceanographic Research Vessel the *Professor Logachev*.

P. 406:
Peter and two US colleagues chat with Anatoli Sagalevitch, head of Mir program, at Copenhagen's Tivoli Gardens.

Other Books
by Peter R. Vogt

Divine Gust of Wind:
Great Trees and Their Afterlives 2023
ages 12–adult

Global scientist and adventurer Peter R. Vogt turns his unbounded curiosity to trees of the Chesapeake Bay, especially the tulip poplar, the tallest deciduous tree in North America. When one majestic specimen topples in a storm, but missing his waterfront home, Vogt masters woodworking skills to carve a dugout canoe modeled on those from mid-Atlantic tidewaters centuries before. The vessel reignites Vogt's wanderlust, and he paddles the balky, 650-pound dugout at storied venues throughout the region, yielding descriptive and comical field notes for the book. *Divine Gust of Wind* focuses on the tulip poplar but also celebrates historic trees throughout America with the wisdom of poets and tree aficionados over centuries.

The Monster Shark's Tooth 2009, 2010, 2020
ages 9–adult

An Arizona boy visits his grandpa near the Calvert Cliffs. They canoe on the Chesapeake and through a cave into the ancient shark-infested Miocene sea.

A Most Mysterious Fossil 2011, 2020
ages 9–adult

The boy returns next year and confronts fossil thieves. He discovers a strange metal object buried in the fossil-rich Calvert Cliffs and with Grandpa's help learns the object's origin.

continued on next page—

What Really Killed the Dinosaurs
—Evidence Found on a Montana Ranch 2022
ages 11–adult

The same boy, now a high-school grad, comes east to work as a Smithsonian museum interpreter. He meets two Montana ranch teens who invite him home to look for dinosaur fossils. The trio discover a strange object with clues (and grandpa's help) to why that asteroid killed the dinos.

Tourmaline's Quest 2022
ages 12–adult

The boy's tomboy cousin visits Grandpa to learn more about her earliest Native American ancestors. She helps him give dugout rides and falls asleep in his beached canoe. She awakes in the world of giant megafauna and sees Indian forebearers. With her grandpa and cousin, she survives the comet fireballs about 13,000 years ago.

Flightship 238 2023
ages 12–adult

A now-world collector buys the diary of a Nazi-brainwashed early teenage boy. The diary details bombing and unintended involvement in a clandestine harrowing late-war escape flight on a giant flying boat. Once in Argentina, he discovers and renounces the evils of Nazism, but the diary ends. The now-world collector and his wife retrace the flight but confront dangers to learn the final destiny of teen, flying boat, and cargo.

www.ingramcontent.com/pod-product-compliance
Lightning Source LLC
Chambersburg PA
CBHW070357130626
46556CB00007B/3197